Examining the COVID Crisis from a Geographical Perspective

This book presents several perspectives on the COVID-19 crisis as it impacted the United States, focusing on policies, practices, and patterns. It considers the relationship between government policies and neo-liberalism, (anti)federalism, economies of scale, and material culture.

The COVID-19 crisis became the primary current event in the United States in March 2020 and continued for several years. In the early days of the crisis, the United States lacked a cohesive, comprehensive approach to combating its spread. As a result, the pandemic was experienced differently in different parts of the United States and at different scales. The chapters in this volume include both quantitative and qualitative explorations of the pandemic as it occurred in the United States. Collectively, they help the reader to better understand this geographically salient issue and provide lessons to learn from so as to improve upon responses to crises in the future.

This book will be of interest to students and researchers of Geography, Sociology, Political Science, and Economics with an interest in United States and the socio-political effects of the COVID-19 pandemic. The chapters in this book were originally published as a special issue of *Geographical Review*.

Sara Beth Keough is Professor of Geography at Saginaw Valley State University, USA. Her research focuses on material cultures of water, and resource-dependency. She is a Fulbright Scholar, the author of *Water, Life, and Profit: Fluid Economies and Cultures in Niamey, Niger* (2019), and Editor of the journal *Material Culture*.

David H. Kaplan is Professor of Geography at Kent State University, USA. His research focuses on nationalism, ethnic segregation, housing, and transportation planning. He is the editor of the *Geographical Review* and has published 14 books and over 65 articles and book chapters.

Examining the COVID Crisis from a Geographical Perspective

Edited by
Sara Beth Keough and David H. Kaplan

Routledge
Taylor & Francis Group

LONDON AND NEW YORK

First published 2023
by Routledge
4 Park Square, Milton Park, Abingdon, Oxon OX14 4RN

and by Routledge
605 Third Avenue, New York, NY 10158

Routledge is an imprint of the Taylor & Francis Group, an informa business

British Library Cataloguing in Publication Data
A catalogue record for this book is available from the British Library

ISBN13: 978-1-032-44713-1 (hbk)
ISBN13: 978-1-032-44714-8 (pbk)
ISBN13: 978-1-003-37354-4 (ebk)

DOI: 10.4324/9781003373544

Typeset in Minion Pro
by Newgen Publishing UK

Publisher's Note
The publisher accepts responsibility for any inconsistencies that may have arisen during the conversion of this book from journal articles to book chapters, namely the inclusion of journal terminology.

Disclaimer
Every effort has been made to contact copyright holders for their permission to reprint material in this book. The publishers would be grateful to hear from any copyright holder who is not here acknowledged and will undertake to rectify any errors or omissions in future editions of this book.

Contents

Citation Information

The chapters in this book were originally published in the journal *Geographical Review*, volume 111, issue 4 (2021). When citing this material, please use the original page numbering for each article, as follows:

Introduction

Introduction to the Special Issue on COVID-19
Sara Beth Keough and David H. Kaplan
Geographical Review, volume 111, issue 4 (2021), pp. 493–495

Chapter 1

The Coronavirus Pandemic and American Neoliberalism
Barney Warf
Geographical Review, volume 111, issue 4 (2021), pp. 496–509

Chapter 2

Anti-Federalist Federalism: American "Populism" and the Spatial Contradictions of US Government in the Time of COVID-19
John Agnew
Geographical Review, volume 111, issue 4 (2021), pp. 510–527

Chapter 3

What are the Impacts of COVID-19 *on Small Businesses in the U.S.? Early Evidence Based on the Largest 50 MSAs*
Qingfang Wang and Wei Kang
Geographical Review, volume 111, issue 4 (2021), pp. 528–557

Chapter 4

Masks and Materiality in the Era of COVID-19
Sara Beth Keough
Geographical Review, volume 111, issue 4 (2021), pp. 558–570

For any permission-related enquiries please visit:
www.tandfonline.com/page/help/permissions

Notes on Contributors

John Agnew, Department of Geography, University of California, Los Angeles, USA.

Gautam Dasarathy, School of Electrical, Computer and Energy Engineering, Arizona State University, USA.

Alexandria Drake, School of Human Evolution and Social Change, Arizona State University, USA.

Samantha Friedman, Department of Sociology, University at Albany SUNY, USA.

Wei Kang, Inland Center for Sustainable Development School of Public Policy, University of North Texas, USA.

David H. Kaplan, Department of Geography, Kent State University, USA.

Sara Beth Keough, Department of Geography, Saginaw Valley State University, USA.

Robert Lattus, Ira A. Fulton Schools of Engineering, Arizona State University, USA.

Jin-Wook Lee, Center for Social and Demographic Analysis, University at Albany SUNY, USA.

Sarbeswar Praharaj, Knowledge Exchange for Resilience, Arizona State University, USA.

Ankith Raaman, Ira A. Fulton Schools of Engineering, Arizona State University, USA.

Akarshan Sajja, Ira A. Fulton Schools of Engineering, Arizona State University, USA.

Patricia Solís, School of Geographical Sciences and Urban Planning, Arizona State University, USA.

Pavan Turaga, School of Arts, Media and Engineering, Arizona State University, USA.

Kevin Jatin Vora, Ira A. Fulton Schools of Engineering, Arizona State University, USA.

Qingfang Wang, School of Public Policy, University of California, Riverside, USA.

Barney Warf, Department of Geography, University of Kansas, USA.

INTRODUCTION: GEOGRAPHICAL PERSPECTIVES ON THE COVID-19 CRISIS

SARA BETH KEOUGH and DAVID H. KAPLAN

$\mathcal{F}$or most in the United States, the realities of the COVID-19 pandemic hit home in March 2020. Businesses closed temporarily, industries moved as many employees as possible into work-from-home situations, schools began virtual education, grocery and food delivery became a safe and convenient way to acquire daily necessities, masks and social distancing became principle means for reducing the virus spread, and people escaped the monotony of their homes for outdoor spaces. In response to the arrival of COVID-19 in the United States, there lacked coherent, cohesive, and comprehensive plan of attack from the federal government. Instead, states and municipalities were left to issue their own directives, using information from a variety of sources. This decentralized method of handling the pandemic resulted in reinterpretations of space, place, practices, policies, and spatial patterns.

One of the most significant shifts in perceptions of place during the pandemic became the distinction between places of risk and places of safety. There was little gray area. Risky places included both those that could be avoided under stay-at-home mandates, such as places of employment for nonessential workers, and those that were deemed vital to everyday life, such as grocery stores or the hospital. Home was generally considered safe from the pandemic, but for those who were living in abusive environments or in situations that lacked access to resources required for a basic standard of living, stay-at-home orders created new problems.

New geographies of power emerged at various scales. In his Forward to the 2013 special issue of *The Geographical Review* on catastrophic geographies, John Agnew writes that "disasters may just happen, but catastrophes are made" (Agnew 2013, 455). Michael Lewis in *The Premonition* (2021) documents how deeply politicized public health has become in the United States, using the Centers for Disease Control and Prevention as an example. The bureaucratic and political convenience which, Lewis claims, has plagued the United States for decades, crippled the country as COVID-19 spread exponentially among the population. In many states, mandates were issued by governors using the power of emergency orders, but these mandates were left to local officials to enforce. As Agnew describes in his article in this issue, the same federal government that failed to promote a comprehensive plan of attack also criticized the governors, like Gretchen Whitmer of Michigan, who issued strict quarantines in their own states. Later in the pandemic, the politics of "opening up," and the speed and pace with which that process should occur, was wrought with controversy. "Safe" and "risky" places deemed so early in the pandemic were perceived differently eight months later, and even more differently after vaccines were widely available. As geography has proven time and again, spaces and places are not stagnant. They change.

Finally, spatial patterns changed as the pandemic wore on. Third spaces, the spaces we interact with that are not home or work, morphed in type and area. In most cases, they shrunk substantially, as bowling allies, movie theaters, shopping malls, sports arenas, and concert halls closed. In other cases, parks, bike paths, forests, and other outdoor spaces became key third spaces, including for those who didn't engage in outdoor activities regularly before. Living in low-density, small cities became an advantage, as the amount of open space could easily accommodate the local population. High-density areas like New York City, where open spaces are disproportionately limited compared to population size, were much more restricted.

This special issue on COVID-19 seeks to explore the pandemic as relevant to geographers and those in related disciplines. The articles in this issue fall into three general categories: policies, practices, and patterns. The articles consider the perspective on the pandemic from the United States, as this country has endured a disproportionately high number of COVID-19 cases and deaths.

At the national scale, understanding the geographic impacts of COVID-19 requires, first, a consideration for the larger geopolitical context. John Agnew and Barney Warf provide this. Warf explores the neoliberalist environment in the United States that resulted in the catastrophic impact of the pandemic. He attributes decades of American exceptionalism, religiosity, anti-intellectualism, and the racial, ethnic, and social inequalities that have created a healthcare crisis in general, and high rates of homelessness and obesity to the poor position the United States was in to handle the pandemic. Agnew examines the spatial contradictions in American federalism, and a lack of action by the central government in particular, which contributed to inequality at various scales. The lack of a comprehensive response to the pandemic at the national level meant that those left to make decisions (states, local governments, law enforcement, and private agencies) differed and floundered in their approach.

As a result of polices (or lack thereof) at the national scale, COVID-19 influenced particular practices at local scales. Qingfang Wang and Wei Kang analyze the impact of the pandemic on small business practices in 50 MSAs and the degree of vulnerability created by policy differences. While they ultimately conclude that neoliberalist policies create greater vulnerability in small businesses, they also found that where greater levels of small business assistance were present, infection rates were lower, and more businesses reported that they expected to recover from the economic slowdown. In other words, policies directly influenced practices and resiliency.

In another example, Sara Beth Keough considers how mask mandates created material cultures of acquisition, display, and disposal of masks during the pandemic. From the State of Michigan perspective, arguably one of the states with the strictest quarantine, mask, and social-distancing policies issued by the governor, Keough finds that the commodity value of masks changed drastically throughout the pandemic, along with the creative ways masks were obtained, worn, and disposed of over time. When masks were mandated, but in short supply, their commodity value was higher and more creative ways for obtaining them could be found. Yet, when production caught up to demand later in the pandemic, the creativity switched from methods of obtaining masks to ways in which individual identities and intentions were expressed through masks.

Finally, the pandemic created new spatial patterns of power and practice as it evolved. Samantha Friedman and Jin-Wook Lee map mortality in New York City neighborhoods by race, ethnicity, and nativity status, showing a hierarchy of impact. Important to note is that though high degrees of vulnerability exist in nonwhite neighborhoods of New York City, the underlying reasons for this variation are complex and uneven, the results of decades of structural and environmental racism, segregation, and disinvestment at multiple scales. At the national level, Patricia Solis and others explore the spatial patterns of governor actions during the pandemic's first wave. They develop a series of models that link COVID-19 mortality to state-level decisions and showing how the patchwork of state data and geographies hinder any sort of unified prediction.

These three categories of policies, practices, and patterns are not mutually exclusive, of course. We encourage you to read all the articles in this issue because, collectively, they create a more holistic, though arguably still incomplete, picture of the early part of the pandemic in the United States. Most authors who considered the federal-level response to the pandemic agree that the unwillingness to act and the lack of a comprehensive plan exacerbated inequalities that already existed for complex reasons. Policies and orders were issued, but the practices that ensued from them produced varied results and levels of success for reasons authors in this issue explain. Personal and political identities and values were expressed materially (through mask wearing, for example) and in reaction to (for or against) state and local policies, and performances by the federal government and its entities.

As of the publication of this special issue, the COVID-19 pandemic is still with us. The Delta variant has forced cases to increase and left the entire world in a morass of uncertainty. Yet the variety of topics and perspectives animating these papers remain. They situate the COVID-19 pandemic as a geographically salient contemporary issue and present opportunities to learn from and improve upon such crises that are likely to arise in the future.

ORCID

Sara Beth Keough ⓘ http://orcid.org/0000-0002-1710-1276

REFERENCES

Agnew, J. 2013. Forward to Special Issue on Catastrophic Geographies. *Geographical Review* 103 (4):455–457. doi:10.1111/j.1931-0846.2013.00012.x
Lewis, J. 2021. *The Premonition: A Pandemic Story*. New York: W. W. Norton.

THE CORONAVIRUS PANDEMIC AND AMERICAN NEOLIBERALISM

BARNEY WARF

ABSTRACT. This essay explores how the coronavirus pandemic has exposed the glaring weaknesses of the American neoliberal order. In early 2021, the U.S., with 4% of the world's population, suffered 20% of the world's COVID-19 cases. The paper examines the factors that made the U.S. uniquely vulnerable to the pandemic: hyper-individualism, high levels of religiosity, ignorance of science, generalized distrust of the government, and the fetishization of the market. These forces produced a society with high levels of income inequality, a tattered social safety net, lack of universal health coverage, a large prison population, and vast numbers of the homeless. The health care crisis gave rise to an economic crisis. Low income communities and people of color are particularly vulnerable. The structural deficiencies of American neoliberalism are exacerbated by the corruption, indifference, and incompetence of the Trump administration, which failed to take the pandemic seriously. Lack of testing, pastors defying social distancing guidelines, forcing low wage workers back to work and the touting of fraudulent cures such as hydroxy-chloroquine illustrate what happens when a conservative society unused to inconvenience encounters an unseen foe that replicates itself with devastating efficiency.

*I*n February 2021, with more than 21 million infected people and a death toll from the COVID-19 pandemic surpassing 500,000 the United States, with 4% of the world's population, endured 1/5 of global coronavirus cases. The actual toll may be much higher. As Young (2020) put it, "How did it come to this? A virus a thousand times smaller than a dust mote has humbled and humiliated the planet's most powerful nation." As other countries managed to "flatten the curve" of new infections, in the U.S. the number of cases climbed steadily. How did the world's only superpower and largest economy, with a formidable set of institutions such as hospitals and research universities, come to such a sad state? The answer lies largely in the uniquely defining features of the country, its conservative political climate, anti-intellectualism, and the incompetence and corruption of the Trump administration (Leonhardt 2020).

This paper approaches the corona pandemic in the United States by emphasizing the unique character of American neoliberalism. Of all the thoroughly neoliberalized societies in the world, the U.S. has arguably most heavily internalized the neoliberal ethos of "free markets," minimal state intervention. It lays out the argument that the U.S. is uniquely vulnerable to the virus because of structural inequalities and weaknesses in the prevailing political and economic order. The paper opens with a brief commentary on the varieties of capitalism literature, which notes that capitalism, and neoliberalism, are geographically differentiated. It links this notion to the long-standing ideology of American exceptionalism. The second section reviews the dimensions of U.S. society that allowed the virus to make such rapid inroads: distrust of the state and fetishization of markets; high degrees of religiosity and anti-intellectualism; individualism; lack of universal health coverage; high levels of obesity and related problems; the world's highest incarceration rate and prison population; homelessness; and systemic racism and racial inequality. The third part examines the Trump administration's response to the pandemic, holding that its incompetence and corruption contributed mightily to high rates of excess deaths. The conclusion summarizes the major themes.

Varieties of Capitalism and American Neoliberalism

Neoliberalism, arguably the most potent political force in the globe today, has received enormous scholarly attention (e.g., England and Ward 2007; Wilson 2017; Balibar et al. 2019). Neoliberalism has been defined and conceptualized in different ways (Birch 2017), but its central elements are well understood: its conservative political agenda; emphasis on "free" markets; and withdrawal of the state from social reproduction. All over the world neoliberalism has shredded public safety nets and accentuated social and spatial inequality. Geographers have also studied this topic (Harvey 2005) and examined how neoliberal principles and practices have reshaped geographies at scales ranging from the local to the global (Peck and Tickell 2002; Peck et al. 2018). As an exchange between Richard Peet (2019) on the one hand and Cahill and Konings (2019) on the other illustrates, neoliberalism is open to interpretation, and means different things in different sociospatial contexts.

In keeping with geographical views of neoliberalism, the "varieties of capitalism" approach to political economy emphasizes institutional differences among countries, acknowledging that capitalism means different things in different spatial contexts (Hall and Soskice 2001; Hancké 2009; Hall 2015). With different labor markets, cultures, historical trajectories, technological environments, and government policies, capitalism in, say, France is quite different from that found in China. This view is valuable in refuting simplistic one-size-fits-all theorizations that ignore profound historical and geographical differences.

In the American context, the varieties of capitalism approach meshes conveniently with the long-standing doctrine of American exceptionalism. American exceptionalism, first proposed by Alexis de Toqueville, focuses on the unique characteristics of American society: the prominent role played by religion; high rates of social mobility; immigration; and so forth. In this view, often widely held by conservatives, the U.S. is a model society unfettered by the normal constraints faced by most countries, a "shining city on the hill" and model of democracy (Lipset 1996; Madsen 1998; Hodgson 2009). The historical path taken by the U.S. led capitalism there to differ markedly from that found in

Western Europe. For example, the broad range of public social services often labeled the "safety net" that typify social democracies such as Sweden or France is relatively impoverished in the U.S. Government programs to help and protect workers, the poor, and the disadvantaged originated largely with the New Deal of the 1930s and the Great Society programs of the 1980s. However, such measures have faced sustained attacks for more than four decades by a Republican Party determined to undo them (Gritter 2018; Dickinson 2019). While the GOP is not alone in its ardent embrace of neoliberalism (the Democratic Party occasionally suffers from this phenomenon too), it does promote neoliberalism as an ideology and set of political practices far more openly than any other party in the Western world (Pühringer and Ötsch 2018), particularly under President Donald Trump (Lachmann 2019). The social safety net has been steadily degraded by design (Abramovitz et al. 2020). Behavioral explanations that blame the poor for poverty, and the fetishization of the market, tend to dominate public discourses about social policy. European-style measures are frequently denounced as "socialism" (Pew Research Center 2019a). As a result, policies and institutions taken for granted in Europe such as workplace protections, unemployment insurance, paid holidays, and unions tend to be relatively weak and underfunded in the U.S. The outcomes of American neoliberalism are not difficult to see. The country has by far the highest level of income inequality among those in the economically developed world (Filauro and Parolin 2019). While unemployment due to the pandemic rose modestly in Europe, in the U.S., by contrast, it soared (Mchugh 2020).

What Explains U.S. Vulnerability to the Coronavirus Pandemic?

Several factors have contributed to the unique vulnerability of the United States in the face of the coronavirus pandemic. In this sense, American exceptionalism has cruelly left the country uniquely vulnerable to the corona pandemic.

Far more than their counterparts in Europe or Japan, Americans tend to distrust their government (Zeleny and Thee-Brenan 2011; Gershtenson and Plane 2015). To some extent this distrust reflects the reality that Washington policy makers are far more responsive to lobbyists and campaign contributors than the voting public. Fueled by a faith in markets and a relentless conservative media assault, many conservatives in particular view the federal government in highly suspicious terms. In its most extreme manifestations this distrust gives rise to conspiracy theories, including ludicrous ones (e.g., Q-Anon). In the context of the coronavirus pandemic, distrust of government led to refusals to follow health guidelines such as wearing face masks and even willingness to take a vaccine if it were created (Bir and Widmar 2020). Especially in the U.S. has the wearing of face masks become highly politicized, with a large proportion of conservatives and Republicans refusing to do so (Smith and Wanless 2020).

Unlike all other industrialized nations, the U.S. exhibits high rates of religiosity. Although organized religion is gradually declining in its popularity (Pew Research Center 2019b), it still remains a formidable force. Americans are far more likely to say that religion is important in their lives, and to go to church or synagogue on a weekly basis, than are Western Europeans (Voas and Chaves 2016; Chaves 2017). Most American Christians pray daily. This religiosity played out with devastating effect during the pandemic. Some conservative preachers denounced the virus as a hoax, claiming that god would protect them (Wilson 2020). Texas-based televangelist Kenneth Copeland claimed he could cure the virus if people watching his show touched their television sets (Niemietz 2020). Some, such as Virginia bishop Gerald Glenn, even caught the virus and died as a result (Brown 2020). More broadly, U.S. religiosity undermined scientific understandings of the virus and rational responses grounded in facts, data, and evidence. As several observers have noted astutely, neoliberalism and religion have long sustained one another (Frank 2001; Moreton 2009; Peters 2018), such as in their mutual naturalization of markets and inequality as well as generalized hostility to progressive forces and ideas. Evangelical Christians in the U.S. have long eagerly espoused free trade and free enterprise as a means of serving god, essentially updating the old Protestant ethic to more contemporary globalized times, an ethos in keeping with the steady commercialization of religion.

The long tradition of American anti-intellectualism (Gore 2007; Jacoby 2008) has undoubtedly contributed to the spread of the coronavirus. Alone among developed countries, the U.S. resists the teaching of evolution (Deniz and Borgerding 2018). Skepticism about anthropogenic climate change is far more widespread in the U.S. than elsewhere (Egan and Mullin 2017). "Experts" are often distrusted as part of some mythical, condescending elite. Increasingly, a widespread distrust of expertise cultivates a climate in which any opinion is as good as any other (Nichols 2017). Under these circumstances, truth simply becomes a matter of perspective. When the decline in objectivity becomes normalized, emotions and affect rise to the fore (Boler and Davis 2018). As waves of fake news and fake science (e.g., climate change denial, anti-vaccination discourses, creationism) wash over the country, it has suffered from what the Rand Corporation has called "truth decay." The very notion of objectivity has come under question and challenged the Enlightenment notion of reason. This phenomenon has been actively abetted by the Republican Party, which has made distrust of science and scientists a major part of its political strategy (Mooney 2006). The danger in this approach is that just when science is most needed, such as the pandemic, it finds the least appeal. Warnings by public health officials may go unheeded. As Krugman (2020a) puts it,

> We're also doing badly because ... there's a longstanding anti-science, anti-expertise streak in American culture—the same streak that makes us uniquely unwilling to accept the reality of evolution or acknowledge the threat of climate change.

American culture also tends to be highly individualistic, a worldview that stresses individual rights, freedoms, and responsibilities and minimizes social obligations (Levin 2017). At times, this trait can be an advantage, contributing, for example, to high levels of innovation and business start-ups. But social problems cannot be handled individually, they require collective action. Many Americans, particularly conservatives, resent *any* limitation on their private behavior. With many people obsessed with a mythologized notion of "freedom," *any* limitation on behavior is regarded as immoral, unfair, and "un-American," and individual sacrifice for the common good becomes demonized (Bunch 2020). Hence, wearing face masks has come to be seen by many conservatives as an infringement of their rights, what Lithwick (2020) calls a "uniquely American pathology." Regular consumers of right wing media, notably Fox News, which have spouted misinformation are particularly likely to dismiss the pandemic as a "liberal hoax," often believe in unfounded conspiracy theories, take pseudo-remedies such as hydroxychloroquine, and refuse to take preventative measures, making them especially vulnerable (Ingraham 2020; Sullivan 2020). Famed conservative radio commentator Rush Limbaugh asserted COVID-19 was "just like the common cold." More broadly, mass ignorance and distrust of science led many Americans to ignore social distancing warnings and engage in reckless behavior, leading to a surge in new cases (Boot 2020).

A prime example of the neoliberal onslaught against social services is the debate over health care coverage. Alone among industrialized countries, the U.S. lacks universal health care insurance. This situation reflects a system in which most health care insurance is provided by private employers, which can be traced back to the wage and price controls imposed during World War II. While the U.S. spends a higher proportion of its GDP on health care than does any other country, its returns for this investment are meager. Medicare and Medicaid provide some assistance to the elderly and low-income families, respectively, but considerable numbers of people lack access to adequate health care. In 2009, 19.5% of the population was uninsured. The passage of the Affordable Health Care (ACA) Act in 2010, popularly known as Obamacare, brought these numbers down: in 2020, only 9% of the population was uninsured. Providing public health care insurance for low-income people, however, drove the Republican Party insane with rage. When Republicans controlled the House of Representatives from 2010 to 2018, they voted to eliminate or replace the ACA 70 times (Riotta 2017). Ultimately, these attempts failed and Obamacare still stands as the law of the land. Nonetheless, even in the midst of the coronavirus pandemic, Republicans continue to wage a legal and political struggle to rescind the act it in a series of legal battles that led all the way to the Supreme Court (Cassady 2020). With widespread layoffs during the coronavirus pandemic, in which at least 40 million people lost their jobs and filed unemployment claims, the number of uninsured rose rapidly (Hellmann 2020).

Another variable contributing to U.S. vulnerability is obesity. Americans are the fattest people in the world (Popkin 2009). In 2019, 42.4% of the populations was technically obese (body mass index of 30 or higher); rates vary among ethnic groups however, including 42.2% for Whites, 49.6% for Blacks, and 44.8% for Latinos (Center for Disease Control and Prevention 2019). Millions more are overweight. Obesity rates contribute significantly to mortality from the coronavirus. Patients with obesity-related preexisting conditions such as heart disease, hypertension, or diabetes are 12 times as likely to die from the coronavirus as those without (Sun 2020). Notably, because obesity rates are highest in the South, the mortality rate there is higher than other parts of the country.

Decades of neoliberal "tough on crime" policies have led the U.S. to have the world's highest incarceration rate and largest prison population (Enns 2016; Egan and Mullin 2017; Alexander 2020). In 2018, the total number of prisoners incarcerated in state and federal prisons was roughly 1.5 million (U.S. Department of Justice 2020). Prisons tend to be overcrowded, unsanitary, and suffer shortages of personal protection equipment such as masks. Many prisoners share communal toilets, showers, and sinks, and social distancing is impossible. Not surprisingly, prisons have become hot spots in the American coronavirus pandemic: the five largest hot spots in the U.S. are behind bars. By December, 2020, more than 275,000 prisoners and staff had become infected and 1,700 had died (Carlisle and Bates 2020). Testing within them remains woefully inadequate. Attempts to mitigate this situation have been met with bureaucratic dysfunction and sluggishness (New York Times 2020). Not surprisingly, the epidemic in prisons has unleashed widespread fear (Williams et al. 2020). Some have experienced riots and hunger strikes in response. Significantly, prisons are not hermetically sealed, self-contained systems: prisoners cycle in and out, and are transferred among them regularly. Corona infections among prisoners also threaten guards, nurses, staff, and visitors, indicating that the contagion there is unlikely to stay behind bars. Some prisons have reduced their populations by a small percent, but they remain distinct threats to the virus's persistence and spread. Like nursing homes—another set of corona hot spots—prisons have been underfunded and overcrowded for years.

The United States also has high levels of homelessness compared to most industrialized countries. In a country with a limited social safety net, deinstitutionalized psychiatric patients, vulnerable veterans, impoverished women with children, runaways and former foster care patients, and people squeezed between jobs that pay too little and housing that costs too much, roughly 500,000 people lack a regular place to sleep (Duffin 2020). In the context of the coronavirus pandemic, homelessness is critical: this is a highly vulnerable population (Lima et al. 2020; Tsai and Wilson 2020). Homeless shelters are often known as "hot spots" of COVID-19 infections (Imbert et al. 2020). Even worse, the recession brought on by the pandemic led to numerous evictions, swelling the ranks of those living on the streets.

One of the defining dimensions of American neoliberalism is racial or ethnic inequality. It has been widely documented that ethnic minorities tend to have lower incomes and wealth than do whites, suffer higher unemployment, and enjoy fewer opportunities, including access to a good education, jobs, and homes (Bruch et al. 2019). This inequality was reflected in the higher rates of coronavirus cases and deaths among the country's minority populations (Pirtle 2020). In May, 2020, Latinos/Hispanics, who comprise 15% of the population, constituted 33% of its corona patients; Black people, who are 13% of the population, comprised 22% (Sun 2020). Compared to Whites, Blacks, Latinos, and Native Americans are 4.5, 4, and 5 times more likely to be hospitalized for the virus, respectively (Sun 2020). The reasons for these ethnic discrepancies in the structural and systemic inequality faced by Latinos and African-Americans (Selden and Berdahl 2020), who tend to be renters living in relatively high-density areas more prone to the virus, including public housing, use public transportation more than do whites, and work in low-income jobs such as retail trade and health care from which there is little refuge from the virus. Black people are incarcerated at much higher rates than whites. In addition, obesity rates among African-Americans tend to be higher than among Whites (CDC 2019), as are rates of smoking, diabetes, hypertension, strokes, and heart disease. The widespread prevalence of such preexisting conditions renders this population more vulnerable. Finally, African-Americans and Latinos also tend to have lower rates of health care insurance than do Whites (Sohn 2017).

THE TRUMP ADMINISTRATION AND THE CORONAVIRUS PANDEMIC

American neoliberalism's inherent weaknesses have been greatly compounded by the incompetence and corruption of the Trump administration. In 2019, before the pandemic began, Trump defunded a pandemic unit at the National Security Council (Reichmann 2020). While epidemiologists sounded the threat, Trump ignored them. Instead, he imposed a travel bans on China, long after the virus had escaped the country, causing a surge of last-minute trips.

Initially, Trump asserted his absolute political control over the pandemic, holding that "When somebody is president of the United States, the authority is total" (Sheth 2020). Later, when confronted with a rising number of infections, he asserted "I don't take responsibility at all" (Dowd 2020). Faced with growing numbers of cases and rising deaths, Trump essentially did nothing, forsaking even the pretense of leadership (Shear et al. 2020). Because the Trump administration effectively abandoned federal attempts to contain the virus, states were left on their own, desperately competing for limited medical supplies.

Rather than invoke the Defense Production Act, which would have mobilized federal resources to procure medical supplies, Trump left it to states and cities to come up with their own strategies, competing with one another for supplies. The result was a patchwork of efforts that reflected local politics and priorities. Some locked down completely, others partially, and yet others not at all. The decentralized governance system that characterizes the U.S. led to lack of unified response; rather, states adopted variety of measures. Because the Trump administration effectively abandoned federal attempts to contain the virus, states were left on their own, desperately competing for limited medical supplies. The result was a patchwork of efforts that reflected local politics and priorities. Some locked down completely, others partially, and yet others not at all. Notably, while the virus claimed most of its victims early in the pandemic in liberal-leaning "blue states" such as New York, it steadily progressed in more conservative ones in the South and Midwest, where refusal to wear face masks became a political statement.

Compounding the problem was Trump's rhetoric of denial. Viewing the pandemic as a threat to his reelection, he consistently downplayed the risk. "It's fading away. It's going to fade away," Trump told Sean Hannity on Fox News (Todisco 2020). He announced that the pandemic would be over by Easter, then called for premature openings of stores and schools. He resorted to racist tropes, calling corona the "China virus" and "Kung flu." Milbank (2020) helpfully collected a number of Trump's sayings about the pandemic, which are worth quoting here:

> The coronavirus is very much under control in the USA. We have it totally under control. I'm not concerned at all. It's one person coming in from China. We pretty much shut it down. It will all work out well. We're in great shape. Doesn't spread widely at all in the United States because of the early actions that myself and my administration took. There's a chance it won't spread. It's something that we have tremendous control over. Looks like by April, you know, in theory, when it gets a little warmer, it miraculously goes away. One day it's like a miracle, it will disappear. Just stay calm. It will go away. The Democrats are politicizing the coronavirus. This is their new hoax. Whatever happens, we're totally prepared. Totally ready. We're rated number one for being prepared. We are so prepared like we never have been prepared. Taking early intense action, we have seen dramatically fewer cases of the virus in the United States. We're very much ahead of everything.

For months, Trump refused to wear a face mask, becoming a model for right-wing science deniers who claimed the whole pandemic was a liberal hoax. He repeatedly touted the drug hydroxychloroquine, even though medical studies showed it did not prevent or cure the virus and could cause serious side effects (Qiu 2020). At one point, Trump even suggested that drinking or injecting bleach might help (Rogers et al. 2020), earning him widespread ridicule. He belittled testing for the virus, saying "I personally think testing is overrated, even though I created the greatest testing machine in history." He falsely claimed that the U.S. had tested more people than the rest of the world combined (BBC News 2020). At a rally in Tulsa, Trump proclaimed "when you do testing to that extent, you're going to find more people; you're going to find more cases. So I said to my people, slow the testing down please" (Abutaleb et al. 2020). When protests against quarantine orders arose in several states, Trump tweeted

"LIBERATE MINNESOTA!," "LIBERATE MICHIGAN!" and "LIBERATE VIRGINIA, and save your great 2nd amendment. It is under siege!", thus conflating public health measures with gun confiscation. Trump and aides repeatedly ignored advice from health care professionals and epidemiologists, including his own advisor Dr. Anthony Fauci, at one point even leading a campaign to discredit him (Mindich 2020). Trump repeatedly called for state economies to reopen, students to return to school, and claimed that large gatherings were safe, all contrary to expert opinion (Baker 2020). In no other country did politicians so openly defy the advice of health experts.

The results of this mishandling and incompetence were catastrophic. By February 2021, with 4 percent of the world's population, the U.S. had more than 20% of its cases, or 21 million, including 480,000 deaths. Daily new cases rose by more than 200,000, and up to 4,000 people per day died of the virus. Each day of the pandemic rivaled the attacks on Pearl Harbor in 1941 or the 2001 terrorist attacks on the World Trade Center in the number of fatalities. Unsurprisingly, Trump's popularity declined in proportion to the rise in cases, with the majority of people disapproving how he handled it; his failure to manage it contributed significantly to his electoral defeat in November, 2020.

Concluding Thoughts

The coronavirus has seeped into every crack and fault line in American society, exposing deep and long-standing class and racial inequalities. A country that should have been well prepared to combat the pandemic instead found itself hamstrung. Combined with a formidably ignorant public fed a daily diet of misinformation from conservative outlets such as Fox News, many people became impervious to scientific advice and opinion. The tradition of American exceptionalism, which long fueled nationalist fantasies, has worked cruelly against its country of origin. Dismissive of facts and expertise and placing their faith in religious leaders, many Americans willfully denied scientific advice such as wearing face masks. Faith in the market and rugged individualism led many to equate public health measures with slavery or socialism, or both. High rates of obesity play a key role in the pandemic's mortality rate. Large numbers of the homeless and prisoners likewise formed vulnerable spawning grounds where the virus spread with ease.

Parallel and deeply intertwined with class hierarchies in the U.S. is a racial hierarchy. Decades, if not centuries, of being denied adequate education, employment, and housing left minority populations uniquely vulnerable. Because minorities tend to live in more crowded conditions than do whites, suffer higher rates of obesity and diabetes, are often forced to go back to work more often even under the threat of the virus, and lack health insurance more frequently, they are at greater risk from infection and suffer a lower ability to overcome it. The result, predictably enough, has been that Latinos and African-

Americans have suffered disproportionately from the coronavirus and died in significant numbers from COVID-19.

All of these predicaments were made significantly worse by the Trump administration, whose response to the pandemic have been nothing short of disastrous. Trump officials grossly underestimated the threat that the virus posed, then failed to take decisive actions to combat it. The incompetence of the Trump government exposed the dangers of undermining the social safety net and disregarding the advice of scientists and health care experts. Krugman (2020b) notes that "Trump's narcissism and solipsism are especially blatant, even flamboyant. But he isn't an outlier; he's more a culmination of the American right's long-term trend toward intellectual degradation. And that degradation, more than Trump's character, is what is leading to vast numbers of unnecessary deaths."

ACKNOWLEDGMENTS

The author thanks two reviewers for their helpful comments and criticisms.

REFERENCES

Abramovitz, M., D. Bhargava, and T. Miles 2020. The US Safety Net Is Degrading by Design. *The Nation* (September 17). https://www.thenation.com/article/economy/coronavirus-social-services/

Abutaleb, Y., T. Telford, and J. Dawsey. 2020. Democrats, public health experts decry Trump for saying he asked officials to slow down coronavirus testing. *Washington Post* (June 21). https://www.washingtonpost.com/politics/2020/06/21/democrats-public-health-experts-decry-trump-saying-he-asked-officials-slow-down-coronavirus-testing/

Alexander, M. 2020. *The New Jim Crow: Mass Incarceration in an Age of Colorblindness*, 2nd ed. New York: The New Press.

Baker, P. 2020. 'Mugged by reality,' Trump finds denial won't stop the pandemic. *New York Times* (July 24). https://www.nytimes.com/2020/07/24/us/politics/coronavirus-trump-denial.html

Balibar, É., S. Brandes, W. Brown, M. Cooper, J. Elyachar, M. Feher, M. Moodie, C. Newfield, D. Plehwe, L. Rofel, and L. Salzinger. 2019. *Mutant Neoliberalism: Market Rule and Political Rupture.* New York: Fordham University Press.

Bir, C., and N. Widmar, 2020. Societal Values and Mask Usage for COVID-19 Control in the US. SSRN 3648562.

Birch, K. 2017. *A Research Agenda for Neoliberalism.* Cheltenham: Edward Elgar.

Boler, M., and E. Davis. 2018. The Affective Politics of the "Post-truth" Era: Feeling Rules and Networked Subjectivity. *Emotion, Space and Society* 27:75–85. doi:10.1016/j.emospa.2018.03.002

Boot, M. 2020. Welcome to the United States of 'Idiocracy'. *Washington Post* (June 30). https://www.washingtonpost.com/opinions/2020/06/30/welcome-united-states-idiocracy/

Brown, L. 2020. Virginia Pastor Who Defiantly Held Church Service Dies of Coronavirus. *New York Post* (April 13). https://nypost.com/2020/04/13/virginia-pastor-who-held-packed-church-service-dies-of-coronavirus/

Bruch, S., A. Rosenthal, and J. Soss. 2019. Unequal Positions: A Relational Approach to Racial Inequality Trends in the US States, 1940–2010. *Social Science History* 43(1):159–184. doi:10.1017/ssh.2018.36

Bunch, W. 2020. America Is Drunk on a Warped Idea of Freedom, and Now It's Killing People. *Philadelphia Inquirer* (June 28). https://www.inquirer.com/opinion/commentary/why-americans-dont-wear-masks-coronavirus-freedom-20200628.html?fbclid=IwAR3anRJZkaM2bQCF5luIADmrSrtKF68r9y9wqd3OVVt6uiyc7U_Q2tztfuE

Cahill, D., and M. Konings. 2019. Book Review: The Strong Emotions Provoked by Talking about Neoliberalism: A Reply to Richard Peet. *Human Geography* 12(2):94–95. doi:10.1177/194277861901200211

Carlisle, C., and J. Bates. 2020. With over 275,000 Infections and 1,700 Deaths, COVID-19 Has Devastated the U.S. Prison and Jail Population. *Time* (December 28). https://time.com/5924211/coronavirus-outbreaks-prisons-jails-vaccines/

Cassady, D. 2020. Supreme Court to Hear Trump Administration's Case against Obamacare after Election. *Forbes* (August 19). https://www.forbes.com/sites/danielcassady/2020/08/19/supreme-court-to-hear-trump-administrations-case-against-obamacare-after-election/#47126ef53844

CDC. 2019. Adult Obesity Facts. https://www.cdc.gov/obesity/data/adult.html#:~:text=The%20prevalence%20of%20obesity%20was%2039.8%25%20and%20affected,Center%20for%20Health%20Statistics%20%28NCHS%29%20data%20brief%20PDF-603KB%5D

Chaves, M. 2017. *American Religion: Contemporary Trends*, 2nd ed. Princeton, NJ: Princeton University Press.

Deniz, D., and L. Borgerding, eds. 2018. *Evolution Education around the Globe*. Dordrecht: Springer.

Dickinson, M. 2019. *Feeding the Crisis: Care and Abandonment in America's Food Safety Net*. Berkeley: University of California Press.

Dowd, M. 2020. Plagued by the President. *New York Times* (March 14). https://www.nytimes.com/2020/03/14/opinion/sunday/trump-coronavirus-national-emergency.html

Duffin, E. 2020. Homelessness in the U.S. - Statistics & Facts. Statista (February 27). https://www.statista.com/topics/5139/homelessness-in-the-us/

Egan, P. J., and Mullin, M. 2017. Climate change: US public opinion. *Annual Review of Political Science* 20: 209–227.

England, K., and K. Ward, eds. 2007. *Neoliberalism: States, Networks, Peoples*. Malden, MA: Blackwell Publication.

Enns, P. 2016. *Incarceration Nation: How the United States Became the Most Punitive Democracy in the World*. Cambridge University Press.

Filauro, S., and Z. Parolin. 2019. Unequal Unions? A Comparative Decomposition of Income Inequality in the European Union and United States. *Journal of European Social Policy* 29 (4):545–563. doi:10.1177/0958928718807332

Frank, T. 2001. *One Market under God: Extreme Capitalism, Market Populism, and the End of Economic Democracy*. Toronto: Anchor Canada.

Gershtenson, J., and D. Plane. 2015. In Government We Distrust: Citizen Skepticism and Democracy in the United States. *The Forum* 13(3):481–505. doi:10.1515/for-2015-0029

Gore, A. 2007. *The Assault on Reason*. New York: Penguin Books.

Gritter, M. 2018. *Republican Presidents and the Safety Net: From Moderation to Backlash*. Rowman & Littlefield.

Hall, P. 2015. Varieties of Capitalism. In *Emerging Trends in the Social and Behavioral Sciences: An Interdisciplinary, Searchable, and Linkable Resource*, 1–15. New York: Wiley.

Hall, P., and D. Soskice, eds. 2001. *Varieties of Capitalism: The Institutional Foundations of Comparative Advantage*. Oxford: Oxford University Press.

Hancké, B., ed. 2009. *Debating Varieties of Capitalism: A Reader*. Oxford: Oxford University Press.

Harvey, D. 2005. *A Brief History of Neoliberalism*. Oxford: Oxford University Press.

Hellmann, J. 2020. Coronavirus Double Whammy: Unemployed and Uninsured. *The Hill* (April 9). https://thehill.com/policy/healthcare/491914-coronavirus-double-whammy-unemployed-and-uninsured

Hodgson, G. 2009. *The Myth of American Exceptionalism*. New Haven: Yale University Press.

Imbert, E., P. Kinley, A. Scarborough, C. Cawley, M. Sankaran, S. Cox, M. Kushel, J. Stoltey, S. Cohen, and J. Fuchs. 2020. Coronavirus Disease 2019 (COVID-19) Outbreak in a San Francisco Homeless Shelter. *Clinical Infectious Diseases*. doi:10.1093/cid/ciaa1071

Ingraham, C. 2020. New Research Explores How Conservative Media Misinformation May Have Intensified the Severity of the Pandemic. *Washington Post* (June 25). https://www.washingtonpost.com/business/2020/06/25/fox-news-hannity-coronavirus-misinformation/?fbclid=IwAR3xH2QzZeAjPnlBw61aLpjrhk-5EmEWXBBNVgizFKv-1RMpBg5tTRonLHw

Jacoby, S. 2008. *The Age of Unreason in a Culture of Lies*. New York: Vintage.

Krugman, P. 2020a. A Plague of Willful Ignorance. *New York Times* (June 22). https://www. nytimes.com/2020/06/22/opinion/coronavirus-trump.html?action=click&module= Opinion&pgtype=Homepage

———. 2020b. Trump and His Infallible Advisers. *New York Times* (May 4). https://www.nytimes. com/2020/05/04/opinion/trump-coronavirus.html

Lachmann, R. 2019. Trump: Authoritarian, Just Another Neoliberal Republican, or Both? *Sociologia, Problemas E Práticas* 89:9–31.

Leonhardt, D. 2020. The Unique U.S. Failure to Control the Virus. *New York Times* (August 6). https://www.nytimes.com/2020/08/06/us/united-states-failure-coronavirus.html?action=click& module=Top%20Stories&pgtype=Homepage

Levin, Y. 2017. *The Fractured Republic: Renewing America's Social Contract in the Age of Individualism.* New York: Basic Books.

Lima, N., R. de Souza, P. Feitosa, J. de Sousa Moreira, C. da Silva, and M. Neto. 2020. People Experiencing Homelessness: Their Potential Exposure to COVID-19. *Psychiatry Research* 288:112945. doi:10.1016/j.psychres.2020.112945

Lipset, S. M. 1996. *American Exceptionalism: A Double-edged Sword.* New York: WW Norton & Company.

Lithwick, D. 2020. Refusing to Wear a Mask Is a Uniquely American Pathology. *Slate* (May 14). https://slate.com/news-and-politics/2020/05/masks-coronavirus-america.html?fbclid= IwAR3gdM-8QHNUA7HX6syUCh4WNsfqInRqkAI6KV1F3aLMXr807sxiSf1xrAM

Madsen, D. 1998. *American Exceptionalism.* Jackson: University Press of Mississippi.

Mchugh, D. 2020. Europe's Safety Net Keeps Millions on Payroll while U.S. Unemployment Soars. *Detroit News* (May 10). https://www.detroitnews.com/story/news/nation/2020/05/10/europe-pandemic-safety-net-us-wage-support-layoffs/111685216/

Milbank, D. 2020. A Trump fireside chat — in his own (unfortunate) words. *Washington Post* (March 31). https://www.washingtonpost.com/opinions/2020/03/31/fireside-chat-trump-should-deliver/

Mindich, D. 2020. Trump's campaign against Fauci ignores the proven path for defeating pandemics. *Washington Post* (July 22). https://www.washingtonpost.com/outlook/2020/07/22/trumps-campaign-against-fauci-ignores-proven-path-defeating-pandemics/

Mooney, C. 2006. *The Republican War on Science.* New York: Basic Books.

Moreton, B. 2009. *To Serve God and Walmart: The Making of Christian Free Enterprise.* Cambridge, MA: Harvard University Press.

New York Times. 2020 (June 25). The Coronavirus Crisis inside Prisons Won't Stay behind Bars. https://www.nytimes.com/2020/06/25/opinion/coronavirus-prisons-compassionate-release. html

News, B. B. C. 2020 (May 15). Coronavirus: President Trump's testing claims fact-checked. https:// www.bbc.com/news/world-us-canada-52493073#:~:text=%22We%20have%20tested%20more% 20than%20all%20countries%20put,is%20more%20tests%20than%20any%20other%20single% 20country.

Nichols, T. 2017. *The Death of Expertise: The Campaign against Established Knowledge and Why It Matters.* Oxford: Oxford University Press.

Niemietz, B. 2020. Televangelist Claims to Cure Coronavirus through Television Sets. *New York Daily News* (March 13). http://www.nydailynews.com/news/national/ny-televangelist-cure-coronavirus-television-sets-20200313-wvkb2aqkwzfvzgu3lzwhw6223u-story.html

Peck, J., N. Brenner, and N. Theodore. 2018. Actually Existing Neoliberalism. In *The Sage Handbook of Neoliberalism* edited by D. Cahill, M. Cooper, M. Konings, and D. Primrose, 3–15. Thousand Oaks, CA: Sage.

Peck, J., and A. Tickell. 2002. Neoliberalizing Space. *Antipode* 34(3):380–404. doi:10.1111/1467-8330.00247

Peet, R. 2019. Book Review: Neoliberalism: Kean Birch. *A Research Agenda for Neoliberalism. Human Geography* 12(2):91–93. doi:10.1177/19427786901200210

Peters, T. 2018. Corporations, Sovereignty and the Religion of Neoliberalism. *Law and Critique* 29 (3):271–292. doi:10.1007/s10978-018-9231-1

Pew Research Center. 2019a. In Their Own Words: Behind Americans' Views of 'Socialism' and 'Capitalism'. (October 7). https://www.pewresearch.org/politics/2019/10/07/in-their-own-words -behind-americans-views-of-socialism-and-capitalism/

_____. 2019b. In U.S., Decline of Christianity Continues at Rapid Pace. https://www.pewforum. org/2019/10/17/in-u-s-decline-of-christianity-continues-at-rapid-pace/

Pirtle, W. 2020. Racial Capitalism: A Fundamental Cause of Novel Coronavirus (COVID-19) Pandemic Inequities in the United States. *Health Education & Behavior* doi:10.1177/ 1090198120922942

Popkin, B. 2009. *The World Is Fat.* New York: Avery.

Pühringer, S., and W. Ötsch. 2018. Neoliberalism and Right-wing Populism: Conceptual Analogies. *Forum for Social Economics* 47(2):193–203. doi:10.1080/07360932.2018.1451765

Qiu, L. 2020. Trump's inaccurate claims on hydroxychloroquine. *New York Times (May 21).* https://www.nytimes.com/2020/05/21/us/politics/trump-fact-check-hydroxychloroquine-coro navirus-.html

Reichmann, D. 2020. Trump Disbanded NSC Pandemic Unit that Experts Had Praised. *ABC News* (March 14). https://abcnews.go.com/Politics/wireStory/trump-disbanded-nsc-pandemic-unit-experts-praised-69594177

Riotta, C. 2017. GOP Aims to Kill Obamacare yet Again after Failing 70 Times. *Newsweek* (July 29). https://www.newsweek.com/gop-health-care-bill-repeal-and-replace-70-failed-attempts-643832

Rogers, K., C. Hauser, A. Yuhas, and M. Haberman. 2020. Trump's Suggestion that Disinfectants Could Be Used to Treat Coronavirus Prompts Aggressive Pushback. *New York Times* (April 24). https://www.nytimes.com/2020/04/24/us/politics/trump-inject-disinfectant-bleach-coronavirus.html

Selden, T., and T. Berdahl. 2020. COVID-19 and Racial/ethnic Disparities in Health Risk, Employment, and Household Composition: Study Examines Potential Explanations for Racial-ethnic Disparities in COVID-19 Hospitalizations and Mortality. *Health Affairs* 39 (9):1624–1632. doi:10.1377/hlthaff.2020.00897

Shear, M., N. Weiland, E. Lipton, M. Haberman, and D. Sanger. 2020. Inside Trump's Failure: The Rush to Abandon Leadership Role on the Virus. *New York Times* (July 20). https://www. nytimes.com/2020/07/18/us/politics/trump-coronavirus-response-failure-leadership.html? action=click&module=Top%20Stories&pgtype=Homepage

Sheth, S. 2020. Trump Falsely Claims that 'When Somebody Is the President of the United States, the Authority Is Total'. *MSN News* (April 14). https://www.msn.com/en-us/news/politics/ trump-falsely-claims-that-when-somebody-is-the-president-of-the-united-states-the-authority -is-total/ar-BB12zQCk

Smith, V., and A. Wanless 2020. Unmasking the Truth: Public Health Experts, the Coronavirus, and the Raucous Marketplace of Ideas. Carnegie Foundation. https://carnegieendowment.org/ files/05_20_Smith_Wanless_Truth.pdf

Sohn, H. 2017. Racial and Ethnic Disparities in Health Insurance Coverage: Dynamics of Gaining and Losing Coverage over the Life-course. *Population Research and Policy Review* 36 (2):181–201. doi:10.1007/s11113-016-9416-y

Sullivan, M. 2020 The Data Is In: Fox News May Have Kept Millions from Taking the Coronavirus Threat Seriously. *Washington Post* (June 28). https://www.washingtonpost.com/lifestyle/ media/the-data-is-in-fox-news-may-have-kept-millions-from-taking-the-coronavirus-threat-seriously/2020/06/26/60d88aa2-b7c3-11ea-a8da-693df3d7674a_story.html

Sun, L. 2020. Patients with Underlying Conditions Were 12 Times as Likely to Die of COVID-19 as Otherwise Healthy People, CDC Finds. *Washington Post* (June 15). https://www.washington post.com/health/2020/06/15/patients-with-underlying-conditions-were-12-times-more-likely-die-covid-19-than-otherwise-healthy-people-cdc-finds/

Todisco, E. 2020. President Trump claims coronavirus will 'fade away' even without vaccine as cases rise in U.S. *New York Times* (June 18). https://people.com/politics/donald-trump-says-coronavirus-fade-away-without-vaccine/

Tsai, J., and M. Wilson. 2020. COVID-19: A Potential Public Health Problem for Homeless Populations. *The Lancet Public Health* 5(4):e186–e187. doi:10.1016/S2468-2667(20)30053-0

U.S. Department of Justice. 2020. Prisoners in 2018. https://www.bjs.gov/content/pub/pdf/p18_ sum.pdf

Voas, D., and M. Chaves. 2016. Is the United States a Counterexample to the Secularization Thesis? *American Journal of Sociology* 121(5):1517–1556. doi:10.1086/684202

Williams, T., L. Seline, and R. Griesbach. 2020. Coronavirus Cases Rise Sharply in Prisons Even as They Plateau Nationwide. *New York Times* (June 17). https://www.nytimes.com/2020/06/16/us/coronavirus-inmates-prisons-jails.html

Wilson, J. 2017. *Neoliberalism*. London: Routledge.

_____ 2020. The Rightwing Christian Preachers in Deep Denial over COVID-19's Danger. *The Guardian* (April 4). https://www.theguardian.com/us-news/2020/apr/04/america-rightwing-christian-preachers-virus-hoax

Young, E. 2020. How the Pandemic Defeated America. *The Atlantic* (September). https://www.theatlantic.com/magazine/archive/2020/09/coronavirus-american-failure/614191/

Zeleny, J., and M. Thee-Brenan 2011. New Poll Finds a Deep Distrust of Government. *New York Times* (October 25). https://www.nytimes.com/2011/10/26/us/politics/poll-finds-anxiety-on-the-economy-fuels-volatility-in-the-2012-race.html

ANTI-FEDERALIST FEDERALISM: AMERICAN "POPULISM" AND THE SPATIAL CONTRADICTIONS OF U.S. GOVERNMENT IN THE TIME OF COVID-19

JOHN AGNEW

ABSTRACT. The US federal government has been widely criticized for its response to the Coronavirus/COVID-19 pandemic. Much of the poor response and outcome has been ascribed to President Trump's personal failure. Yet more importantly this failure has been of the US governmental system. More specifically, the role of the federal government in fashioning nationwide policies across a range of areas, including public health, has been crippled by an anti-federalist ideology and the institutional inertia it has created. Ordinarily, one would think that the federal government would be empowered by a self-defined "nationalist" or right-wing populist in the White House. But rather than command and coordination across tiers of government, the states have been left to cope as best they can without much of anything in terms of coherent and consistent national/federal leadership. The recent efflorescence of anti-federalist ideology has roots going back to the 1980s. The pandemic has exposed the distortion of the once well-established polyphonic practices of historic US federalism by a now institutionalized dualist vision of federalism that has sadly become the leitmotif of failed US governance in the pandemic.

"Disasters may just happen, but catastrophes are made" (Agnew 2013, 455).

Pandemics are by definition global in character and spread from place to place through travel and community spread. They are a test for different governmental systems and the geopolitical-territorial arrangements upon which they rest (Horton 2020). Absent much in the way of effective global governance, national-level governments typically have the best resources and expertise to limit spread and manage healthcare as the disease spreads across their territories. Simply leaving management to lower tiers of government can create major problems when they adopt different testing and mitigation strategies and do not have adequate resources and expertise to institute them. But centralization is also problematic insofar as it presupposes a one-size fits all strategy that fails to account for regional and local specificities in susceptibility and resources to manage a major health emergency. The basic problem is something brought to light many years ago by Mark Twain when, in *Tom Sawyer Abroad* (1894), he alerted us to the fact that coloring in territories on a map as totally distinctive domains is utterly misleading, whether these be the states of the United States he had in mind or the self-evident nation-states of the world political map (Agnew 2019). Viruses do not spread or succumb to suppression solely on a territorial basis. That also goes for a host of other things that require regulation and management. Given that we are stuck with multi-tier territorial governance, how best can that be put to work on our collective behalf?

In the United States there has been much controversy over the problematic role of the federal government in the COVID-19 pandemic. The US has had one of the highest death rates relative to population size worldwide and one of the most incoherent and inconsistent national governmental responses in terms of public health policies and financial support to the tiers of government (state and local) at which policy has been administered (e.g., Beaubien 2020; Lerner et al. 2020). The lack of any national plan for dealing with the pandemic has been particularly obvious from the outset (e.g., Haffajee and Mello 2020). Much of the critique has focused on the performance of President Donald Trump; from his months-long dismissal of the dangers posed by the pandemic to his chaotic administrative approach to the challenges posed by the spread of the virus and his politicization of the pandemic for electoral purposes (e.g., Kristof 2020). Trump has certainly been more part of the problem than any sort of cure. But the pandemic exposes rather deeper structural flaws in the US system of government than just the managerial, psychological and intellectual flaws displayed by the person who has just happened to occupy the office of President at this trying time (see, e.g., Trump 2020; Woodward 2020). So much of what has gone for political debate in the United States since the 1980s has been over the *size of government* when it should have been over the *quality of governance* with respect to what the federal government should and can do. The question is what the government must do that other actors cannot, and also how one can coordinate between them and across tiers of government from the federal to the local (Agnew 2011). This has been a major question in political geography down the years to which the literature has tended to argue for coordination rather than competition or a rigid division of labor between tiers (e.g., Dikshit 1976; Markusen 1994; Van der Wusten 2017).

The major danger inherent in any federalism lies in not allocating "sufficient powers to their general governments to deal with modern economic [and other] crises" (Wheare 1963, 244). Crucially in the United States, a federal Senate biased to favor low-density rural parts of the country and an Electoral College that reinforces this bias in presidential elections have led to a mobilization against the sort of relatively powerful and redistributive central government that Founding Father Alexander Hamilton favored (Rodden 2019; Deaton 2020). A minority of the population has thus come to have a major influence in reviving the radically dualist (states versus federal) vision of US federalism (Milhiser 2021). This vision was the one that had defined the so-called Philadelphian system of weak federal government and strong states that prevailed before the Civil War (Deudney 1995). It was fully revived beginning in the 1980s. This has led to the present impasse where the federal government and its proper functions have been systematically neglected or actively undermined to the detriment of the system as a whole (Mishra 2020).

More specifically, there is a fundamental contradiction between a President like Donald Trump elected on a national-populist basis and the reality of a US governmental system that since the 1980s has been increasingly anti-federalist in its legislative and executive preferences for privatization. It has also favored "small government" and is thus immune to any sort of forward-looking role for the federal government in domestic policy. The manifest federal-government failures in managing the COVID-19 pandemic in the US are the outcome of this contradiction. But it is the federal-state-local nexus that has been more central to the mismanagement than the bureaucratic failures of the federal government per se, even though they have long played a role in skepticism about the performance of the US federal government (e.g., Schuck 2014). To the extent that discussion of the federal failure in the pandemic has extended beyond Trump and his coterie it has been to the specific failings of such federal agencies as the CDC, NIH, and the FDA (e.g., Bandler et al. 2020; Piller 2020). Notwithstanding their importance, I want to argue that the bigger problem has been a geographical one: the revival of a dualist model of federalism that does not match the multiple geographies of power at work in the world and across the United States.

"Revival " of dualism is the operative phrase. A New Deal coalition dominated by the Democrats but subsequently supported also by many Republicans prevailed from the 1930s down until the 1970s. It knitted the country together across a number of significant economic and cultural divides. In the early 1960s the United States was widely viewed, inside and outside the country, as having an exemplary "civic culture" compared to many other countries in terms of trust in government at all levels and in a sense of popular collective political efficacy (e.g. Almond and Verba 1963). But it unraveled beginning in the 1980s though the 1990s with the emergence of major urban-rural and sectional differences on a range of policy issues not least the role of the federal government in the American economy and society (Mellow 2008). Rather than acting as partners, the states and the federal government were increasingly seen as opponents. During the Progressive era in the early twentieth century and following the 1930s New Deal down until the early 1970s, a federalist rebalancing as a result of social mobilizations had led to a more efficient and redistributive national government. Given the ambiguities surrounding the US constitutional compact, disputes between federalists and anti-federalists, originalists and proponents of a living constitution, and so on, the relative legitimacy of the powers of the various tiers of government (and delegation of executive powers) is ideologically forever up for grabs (e.g., Ollmann and Birnbaum 1990; Balkin and Siegel 2009; Mortenson and Bagley 2020). But Trump's rhetorical dictatorial style with support from minions like Attorney General William Barr devoted to a "powerful" executive vis-à-vis the legislative and judicial branches should not be confused with an empowered and effective federal government tout court (e.g., Schwartz 2020; Falconer 2020). This is central to the paradox of Trump's presidency.

Arguably, testing is crucial in managing a pandemic. The highest tier of government could be expected to take a leading role in coordinating across all lower-level jurisdictions in this respect at least. But as late as September 2020 the federal government was acting as if the pandemic were past and still leaving the states and municipalities to cope largely on their own (e.g., Weiner and Helderman 2020). There was still no national testing program by 12 June 2020, five months into the pandemic. Things came together tentatively on testing nationwide only on 22 June. But testing then became a topic that Trump could use to underplay the seriousness of the pandemic. Too much testing, according to Trump, was painting too dire a picture. As a result, in late June 2020 Trump even suggested that testing ought to be slowed down to make the pandemic look better than it was (Cohen 2020). In late June federal funding was pulled from testing sites in five states undergoing major spurts in cases and hospitalizations (LA Times Editorial 2020). At the same time, national testing capacity was still much less than needed to respond to the spread of the pandemic across the country (Madrigal and Meyer 2020b). Even as Trump continued to boast about "our great testing program," as of early July many cities still could not test all those they needed to in order to trace and isolate spreaders, suggesting how deluded the president was about the empirical reality the country faced (Weiner 2020). Contact tracing, absolutely key to suppressing the spread of infection, turned out to be a total debacle, despite the best efforts of some state governments (Steinhauer and Goodnough 2020; Khazan 2020). In early August a group of seven states (Maryland, Virginia, Michigan, Ohio, Louisiana, Massachusetts, and North Carolina) *abandoned* the federal government for their own consortium in quickening the pace and scope of testing (Baker and Court 2020). Trump's delusion about the unconstrained course of the pandemic fed into his inability to see that *sustainably* opening up the economy depended on dealing with the pandemic, not denying its existence (Bassett and Linos 2020). The US strategy, if that word is appropriate, was exactly backwards: instead of conquering the virus first and then opening up, opening up was seen as the priority much too early and then the virus rebounded. As the *Economist* (2020) wrote: it was like a hospital investing "in palliative care while abolishing the oncology department."

After providing an overview of the spatial uncertainties of multi-tier governance revealed by the pandemic both in the US and elsewhere, I turn first to the case for a putatively national-populist leader such as Trump, how he has campaigned and ruled, and then briefly to the paradox of the geopolitical framing at the center of Trump's appeal: the opposition between the national and the global. Suggesting that Trump's performance during the pandemic cannot be understood entirely in these terms at all, I turn to the longer term institutional imbalances in the US federal system that have hobbled response to the pandemic. Of particular importance I claim has been the lack of coordination across the tiers of government from the federal through the state to the local that is a by-product of the anti-federalist perspective on US federalism that has become dominant politically in the US since the 1980s. The dualistic view of federalism (federalist versus anti-federalist) has undermined the possibility of a more polyphonic practice that would have led to better management of the pandemic. Just blaming Trump, therefore, has been to miss noting a more systematic institutional failure.

SPATIAL UNCERTAINTIES OF CONTEMPORARY GOVERNANCE: DUALISM VERSUS POLYPHONY IN FEDERAL GOVERNANCE

In Europe regional-level politicians and big-city mayors have been at odds with national governments over what policies to follow in order to suppress and/or mitigate the COVID-19 pandemic (e.g., Hall et al. 2020). In the United States state governors have clashed with the President and his administration and local officials with state and federal ones over public health measures such as face-masks, physical distancing, and quarantine rules. At one point the President even encouraged armed supporters of his to "liberate" states such as Michigan run by Democratic governors that he considered his adversaries (e.g., Cook and Diamond 2020; Edelman 2020). In this context there has been widespread disagreement about the relative merits of more or less decentralization in managing a crisis of such proportions as the COVID-19 pandemic. After an initial disastrous response to the first outbreak in Wuhan, the highly centralized authoritarian Chinese government brought the pandemic under control in its territory relatively quickly and effectively. At first, but with later problems, the federal German system produced a relatively positive outcome (Studemann 2020). The asymmetric devolution in the UK and the US federal system both produced relatively poor outcomes in terms of cases, hospitalizations, and deaths. Cherry-picking your cases, however, can lead to whatever conclusion you wish.

Political debates in many countries about the relative balance of powers between different tiers of government tend to discuss them in terms of neat divisions between competencies and autonomies exercised at different levels: national, regional, and municipal (see Treisman 2007; Agnew 2011). From this viewpoint, a technical process of matching functions to levels on the basis of externality effects from and popular demands for different public goods and services will lead automatically to a clear and demonstrable geographical separation of powers. Historically, however, different jurisdictional levels have large areas of joint or concurrent powers or what can be termed "polyphony." In the United States, for example, federal and state laws frequently regulate the very same goods and conduct, from drug trafficking to education, gun control, and bond trading. There is no tight combination of territorial level of governance and particular good or service for a wide range of goods and services. There is a constant *practical* struggle between tiers of government over powers in relation to numerous issue-areas. Even what might seem to be areas "settled" at one level, say immigration regulation and foreign policy actions, have become subject to cross-jurisdictional dispute and coordination. Witness popular attempts in California to impose immigration regulations (usually defined in a dualist perspective as a purely "federal" function) and Massachusetts's so-called Burma Law banning all state agencies from signing contracts with companies active in Burma (Myanmar) because of the Burmese government's history of human rights violations (e.g., Guay 2000; Paul 2002). That the US federal courts have turned these measures back is only an indication of how extreme any "sharing" of functions of this type would be. Many other functions, however, cannot be limited solely to one tier of government yet are typically viewed these days as if they should be.

In practice, therefore, governance is rarely if ever about exclusive and non-overlapping spheres of authority neatly divided between geographic scales or tiers of government. The term "polyphony," coined in this context by the American constitutional lawyer Robert Schapiro (2005-6, 2009) to refer to the interaction, competition, and coordination between state and federal powers in the United States, captures much better the actual practices of fuzzy definition, competition, and antagonism that typically inform all attempts at managing power within multi-tier governance. Schapiro sees the dualist vision in the United States as having had twin roots. One is the economic argument of the state as a "firm" or private corporation within a system in which the federal government then operates as an agent of antitrust. In this market model, popular with conservative lawyers and with conservative-dominated US Supreme Courts since the late nineteenth century, federalism is essentially an exercise in line drawing between two tiers of government. Gerald Frug (Frug 2001; Barron and Frug 2006) has pointed to problems with the historical accuracy of this account and specifically to the analogy that inspires it, not least to the fact that local government in the US is increasingly "defensive" rather than reflective of "true" autonomy, and this is so largely because US federal courts have consistently favored private corporations (i.e., businesses) and their expansive operations over public ones such as municipalities.

The other root of dualism lies in the normative republican model that is often held to have motivated the founders of the United States and, more particularly, inspired the writing of the US Constitution. From this perspective, a rigid specification and separation of functions between tiers is held to promote various vague but rhetorically powerful goals: "efficient and responsive government, participatory self-government, and protection against tyranny" (Schapiro 2005-6, 248). The dualist view's continuing ideological attraction, therefore, is longstanding and continues to draw adherence because it is deeply connected to common narratives about US history and US constitutional exceptionalism (Lim 2014). Obviously, the US federal experience sets it apart from countries such as France, Spain, Italy, and Britain, to name just a few, with very different governmental histories and dominant regime ideologies on the continuum from unitary to federal systems. But the distinction that Schapiro draws between dualist and polyphonic perspectives on multi-tier governance is a useful one that can be applied more widely. The tension between them reflects a more realistic grasp of contemporary geographies of power than simply accepting the older dualistic one as an inevitable fait accompli.

Why does the polyphonic perspective make more sense? There are at least three ways of thinking about how power is related to space with respect to political institutions: bounded territories, networked flows, and topological ties (Allen 2009). Typically, we think of governance almost entirely in terms of the first and arguably, for much of human history, this has made considerable sense. Territories contain power in the sense of being "tiered" units of space in which services, for example, are supposedly provided on an equal basis within the unit as a whole or differentially as a result of conscious political decisions within the various parts. The "scaling up" of power and the workings of power more generally, however, often, and increasingly, involves networked flows across territories including across their borders. The need for higher tiers of government can reflect the sense that only at a larger territorial scale can networked flows be managed or regulated. Relations of connection, though, are not always simply topographical, reaching across or bounding concrete spaces. They are also topological: gaps between "here" and "there" are increasingly temporal moments (across the Internet, for example) rather than distanced connectivities. Indeed, powers of reach are potentially beyond territorial containment.

It is important to identify these multiple spatialities of power because at least part of the issue with contemporary questions of territorial governance is the degree to which the externalities emanating from the second two spatial modalities can be captured or entrained within any sort of territorial framework (Agnew 2018). As yet, means of managing flows and more diffuse relations of connection outside of any sort of territorial reference remain radically under-developed. Plausibly, the financial products at the heart of the financial/economic collapse of 2008–9 flowed in real time between networked nodes around the world but were increasingly free of territorial regulation simply because the products and their agents defied the territorial imagination upon which regulation has been largely based. Likewise, our conventional thinking about how to challenge shadowy terrorist networks remains to a high degree trapped within a territorial imagination that can only work on a state-by-state basis rather than directly adapting to the modus operandi of the groups in question. Pandemics and how they spread are not all that different either (Horton 2020).

So, governance is no longer simply a question of matching "functions" to the most appropriate level of territorial resolution (local, metropolitan, regional, national, etc.), as for example in Mann's (1984) classic account of the territorial origins of state autonomy, but also of adapting territorial modes of governance to more complex spatial modalities of power. Arguably, the contemporary world is more pluralistic in its spatialities of power than are available means for managing them politically. But this is also a contributory factor to why any sort of strict division of powers territorially is increasingly problematic in a world where power is not even contingently always divisible territorially (Agnew 1994, 2019). The pandemic has been exhibit A in showing how much practice of a coordinated and overlapping rather than mutually exclusive and divided territorial model of governance would favor better management. Arguably, for all their problems, the cases of Germany and Australia suggest how much a polyphonic federalism can serve to mitigate the disastrous effects of a pandemic as opposed to the either/or opposition between decentralization and centralization that has tended to prevail latterly in the United States.

Donald Trump and National-Populism

Donald Trump campaigned for the US presidency in 2016, unlike previous Republican candidates for that office, on an openly populist platform (Agnew and Shin 2019). His central claim, emblazoned on the baseball hats of his supporters, was "To Make America Great Again." Following on a two-term first-time African-American President, whom Trump had personally insulted and run down from before the 2008 election, including being the primary source of the charge that Obama was an illegitimate president because he had not been born in the US, this slogan was not hard to decode. Indeed, since his election much of what Trump has done has been to undo what Obama had done with respect of social, healthcare and environmental regulation (John 2020). Apart from that, Trump has followed recent Republican orthodoxy on slashing the federal income tax on high-payers and appointing ultra-conservative judges to the federal courts. In the 2016 election campaign, however, more than these initiatives, Trump emphasized "toughness" in "bringing back" jobs in manufacturing that had somehow been stolen by "China" (not a word about the role of US multi-national businesses in this) and building a wall with Mexico (that Mexico would pay for) to keep out the "illegals" that he spent much energy on the campaign trail decrying for their criminality and threat to the racial composition of the country. The entire thrust of Trump's public persona has been to present himself as a national savior with a very clear sense that those he desires to see exalted after the Obama years is the largely elderly white demographic that he appealed to support him in 2016. Since arriving in office he made no attempt to portray himself as a president of the entire country, *only* of those who display loyalty to him (Wehner 2020).

Attacking the "mainstream media" (particularly so-called quality newspapers and television news that report in a fact-driven rather than ideological way) played a vital part in establishing Trump in his prophetic role as leader. The media must be discredited to undermine the empirical truth in which they claim to trade. Steve Bannon, Trump's house theoretician, insisted that the imperative is to dominate the conversation, not to engage in a battle of ideas: "The Democrats don't matter," he says. "The real opposition is the media, and the way to deal with them is to flood the zone with shit" (quoted in Thornhill 2018). Trump has thus appealed to a cultural vein in American society that is suspicious of specialist knowledge and the notion of objective truth. His most important stock-in-trade is to accuse all and sundry who are not loyal to the lies and fabrications he espouses of trading in "fake news." This is how he appeals above all to "his people" or "base."

Say what you will about him, but Trump has been a political genius in managing to conquer a Republican Party that initially was allergic to his appeal, particularly on economic issues such as trade barriers, and in his consistently receiving since his 2016 election support in opinion polls of around 80% percent or so of self-identified Republicans polled through October 2020. In the 2020 presidential election, even while losing nationally and in the Electoral College, he still received around 74 million votes. So, even in the face of a dismal record of mismanaging the early warnings of the coronavirus pandemic, Trump still retained significant popular support (Gabriel and Lerer 2020). His reservoir of support among Republican Party voters is based on a "fealty, a visceral and emotional attachment" that is still triggered by his open displays of nativism and attachment to a nostalgic vision of an America that had been "lost" (quoted in Waldmeir 2020). Indeed, in parts of rural/small town America, his supporters were already prepared to blame their globalist co-nationals who travel abroad for the virus coming into their America (Kilgore 2020). That he rhetorically continued by and large to demonize his political opponents and rewrite his own history in relation to the pandemic shows how much he had not changed operationally even as the challenges he faced were no longer those of his own invention, like the Ukraine imbroglio that led to his first impeachment, but something that would test even the best of leaders (e.g., Baker 2020; Bump 2020; Parker and Rucker 2020).

Even in the face of the most significant challenge facing a US president in a generation he remained focused on his reelection in November 2020 rather than dealing with the crisis at hand. Populism always seems to privilege campaigning over governing, not least because its main tenets, beyond claiming "the people" as its leitmotif, are riling up anger and resentments rather than pursuing rational policy goals or good governance per se (Agnew and Shin 2019). Trump's performance in a prime-time speech about the pandemic on 11 March 2020 as he struck a "starkly militaristic and nationalistic tone" while the country was being radically upended by what he termed a "foreign virus," as if it were not already abroad in the land, was widely panned by critics (Glasser 2020). But it probably resonated positively with those he wished to mobilize for the November 2020 presidential election. Because the pandemic had started in China, even as their favorite son was off playing golf at one of his own resorts, and rallying his base in rambling soliloquys rather than preparing administratively for the pandemic no longer just on the horizon, he was not held responsible.

Key to the entire geopolitical framing that brought Trump to the White House was the discursive opposition between globalism (and globalists) on the one hand and nationalists favoring the people and its national state on the other. The fusion of an idealized people with the national state is by no means alien to American political development (Peel 2018). This framing was the one suggested by Steve Bannon in which rather than pitching himself as the agent of Wall Street and as a business-as-usual Republican, the only way Trump could win in 2016 was in bringing into national electoral politics people alienated from both of the dominant parties by the lackluster performance of the US domestic manufacturing sector and slumping median household incomes since the 1990s. In turn, the best way to do this was to criticize the liberal global order and talk about reestablishing a territorial sovereignty over borders and the economy that had been lost with the latest round of globalization since the 1980s. Imposing tariffs and opposing international trade agreements were the main strategies used to pursue these goals, even as massive tax cuts widened the federal government fiscal deficit that could only be financed by foreign sales of US treasury bonds.

At the same time, of course, Trump was himself very clearly a globalizer with his foreign investments in hotels and golf courses. His cover on this was to paint himself as an American everyman down to how he spoke and what he ate. This is a typical move on the part of right-wing populists everywhere. His business "successes" therefore (notwithstanding a long history of bankruptcies and questionable loans) could be viewed as evidence of his managerial intelligence even as he had to overcome the disability of being just another everyman. As a neo-patrimonial figure dispensing favors to his subjects/people, Trump would reward his supporters through punishing foreign interests and by channeling federal resources and tax-favored capitalist investment to their benighted communities (Riley 2017). This self-presentation met with enormous success among a significant portion of the electorate concentrated largely in southern and western states but with enough strength in what turned out to be the crucial states (given the indirect nature of US presidential elections through the Electoral College) of Michigan, Pennsylvania, and Wisconsin to give him a victory in 2016 even as he failed to achieve a majority of the national vote. As a caveat, I should note that both he and his opponent, Hillary Clinton, had the largest negative approval ratings of any presidential candidates since polls had asked the question (Agnew and Shin 2019).

The claim to a national people as the primary constituency, even though we know that Trump supporters tend to be a very particular demographic-cultural grouping, is central to the entire populist rationale. In the US case this rests first and foremost on ideas about the founding groups and their racial-ethnic profiles.

These, of course, are people of primarily Western European ancestry like Trump himself. When Trump first declared his presidential candidacy, as he descended the escalator at Trump Tower in New York City in 2015, he made his case centrally by declaring what he was *against*, in the case at hand, Mexican immigrants crossing the southern border of the US and defiling the national space by their very presence, to which his answer would be to build a wall and otherwise close off the United States from the rest of the world as best he could. This was Trump's national-populist promise.

The Retreat of the Federal Government Since the 1980s

While representing "his" people, presumably a national constituency at least in theory, Donald Trump has also been heir to a set of ideological positions that have been to a considerable extent contradictory to his national-populist claim. These were apparent in his 2016 campaign but became glaringly obvious in the years in office. Certainly, hostility to professional expertise and science and disdain for disinterested journalism are often fundamental components of right-wing populism (e.g., Gerson 2020). But in the contemporary United States they are frequently connected popularly to government. President Ronald Reagan famously announced in his inaugural address as President of the United States that "Government is not the solution to our problem, government is the problem." Reagan did not so much have professional expertise in mind. But he certainly wished to trim and limit the role of the federal government. He opened the door to doubts about the very idea of the "public interest" and disinterested pursuit of objective knowledge. This reflected a longstanding political current in the United States increasingly dominant since the 1960s in the Republican Party suspicious of the expanded role of the federal government in enforcing regulations on business and civil rights on the population at large. At the same time, however, as noted by Janen Ganesh (2020): "Republicans seem to mistake the public's cynicism about 'government' in the abstract with indifference to actual services and fiscal transfers."

The very term Federalist was redefined to mean the exact opposite of what it meant to the writers of the US Constitution (e.g., Agnew 2005, pp. 102–18; Edling 2003). Thus the right-wing Federalist Society is in fact largely anti-federalist in orientation, belittling and undermining the roles of the federal government that Madison and Hamilton had championed (see, e.g., Ketcham 1986; Hamilton et al. 2014 [1788]). Trump has picked up on this truly anti-federalist viewpoint in his attacks on the purpose and expertise of the federal government tout court and in relation to the experts in government agencies such as the EPA, the Department of the Interior, the Department of State, the FBI and the Department of Justice, and the Department of Defense. Shrinking the role of the federal government thus fulfilled the view that markets and maybe local governments were always better than the central government and that there is no such thing as the public interest (e.g., Frank 2008; Brown 2019).

The Reagan years marked the beginning of what has been called the neoliberal assault on the role of the federal government in managing the US economy and providing for the expansion and protection of fundamental civil rights. From the neoliberal perspective, the best government is that which does least, except insofar as it favors privatized solutions and capitalist interests over public institutions. In practice this was to declare an open season "to strip-mine public assets for the benefit of private interests" (Packer 2020). It was an organized reaction against the so-called liberal-Keynesian view that governments should use fiscal policy, government spending and tax increases to stimulate demand during economic downturns. In its place neo-liberalism variously encouraged monetary as opposed to fiscal policy and tax cuts, particularly on the wealthy and business, as supply-side stimulus. It also preferred private to public provision even of goods, such as healthcare, that most people might reasonably regard as better made available on a public basis. Trump certainly governed in this neoliberal vein (Packer 2020).

At the same time, Trump inherited and cultivated the anti-federalist vote that came out of the civil rights struggles of the 1960s and led to the Republican strategy since Richard Nixon of hunting for white voters in the US South (Maxwell and Shields 2019). From this viewpoint, the federal government represents both the hated "Union" that won the Civil War and the imposition on the South of norms and regulations that do not fit their "heritage." This heritage, as Maxwell and Shields (2019) brilliantly deconstruct, consists of an amalgam of white racism, patriarchy, and religious zealotry used to justify the other two. In this construction, the "Deep State" to which Trump frequently refers, typically associated with right-wing conspiracy theories, is not the bugaboo that libertarians might associate with limiting access to certain calibers of guns or imposing vaccinations, although these can be present too, but more the sense of a national-level government that imposes rules such as affirmative action, restricts local law enforcement, enforces environmental regulations, and insists on the basic equality of all citizens in the eyes of the law. With more than a nod to a Confederate imaginary of the United States, Reagan in his day often used the locution "these" United States to emphasize the sovereignty of the states against that of the federal government. Trump's recourse to the rhetoric of culture war over abortion, gay rights, immigrant undermining of American "culture" and so on all are designed to appeal to a constituency that sees the federal government (particularly the federal judiciary) as useful only in the negative sense of restricting its enforcement powers in the jurisdictions where they live rather in terms of an affirmative role in providing public goods and services on an equal basis nationwide.

The net effect of these two trends toward an anti-federalist conception of the federal government has been to produce an increasingly paralyzed and ineffective national government apparatus. Beyond this, however, the impact has also been to invest in a sort of "Darwinian" federalism in which the states are essentially left to their own devices without the necessary support and leadership of the federal government (Cook and Diamond 2020). This federal failure was on full display in relation to the crisis spawned by the spread of COVID-19. If rather than the either/or logic of the dualist vision a practical polyphony had been at work, the states and the federal government would have operated as alternative and coordinating power centers. So, rather than asking if some function, like public health or pandemic management, "belongs" to one tier or another, we should ask how overlapping and coordinating power operates and can actually improve how some issues are addressed. What is of most use in the distinction between dualism and polyphony that Schapiro (2005-6) makes is that it draws attention to values of plurality, dialogue and redundancy in the latter over against those of uniformity, finality, and hierarchical accountability associated with the dualist vision. The point is not to ennoble the polyphonic alternative conception normatively so much as see it as methodologically more useful in terms of the workings of multi-tier governance in practice, for example in relation to managing a pandemic. In the crunch, Australian and German federalism seemed to exhibit more of this polyphony than did that of the United States.

THE SPATIAL PARADOX OF TRUMP'S "POPULISM" AND THE COVID-19 PANDEMIC

So, at the same time Donald Trump has appealed to a conception of a tightly walled and contained national-territorial homeland, he is also heir to a weakened federal government and federal system that is the outcome of years of systematic degradation at the hands of anti-federalists of several types. On the first count there have been the years of systematic underfunding of national agencies devoted to health and welfare (e.g., Himmelstein and Woolhandler 2016). This reflects a bias against public funding (and federal taxation) and a preference for private initiatives with limited regulatory controls. On the second count there has been a trend to leave all sorts of issues, such as health care finance and provision, entirely to the states and localities (Kettl 2020). This reflects in part the view of the federal government as a usurper of local "heritage" and traditions, and the dangerous enforcer of equal citizenship and rights. It is not so much that a case cannot be made for effective local and regional democracy but that the federal role as a coordinator and manager has been systematically sidelined because of an anti-federalist ideology that has completely vitiated Trump's claim to represent an idealized national-people walled off from the rest of the world.

Trump himself weakened the federal government in very specific ways since arriving in office in 2017, not least in relation to public health management. Trump's neo-patrimonial promises referred to previously as important to his 2016 campaign for President were largely forgotten. The promised investments in national infrastructure and in replanting manufacturing industry came to little or nothing. Would these have finally mattered to his reelection? Perhaps not, but more importantly he made numerous promises to address the COVID-19 crisis *practically* rather than just rhetorically but without much delivery that may well have come back to haunt him in November 2020 (Drezner 2020). His main achievements in office were a giant income tax cut for the wealthy and business in 2017 and the appointment of numerous ultra-conservative judges to the federal courts. Even as he continued with his populist-nationalist rhetoric, Trump systematically degraded the functioning of the US federal government (e.g., Bergen 2019; Rucker and Leonnig 2020). Federal government departments had thin or acting leadership for long periods. Many of the political appointees running their agencies were utterly incompetent for or opposed to the charges they received. Regulations and rules were rolled back across the board from education to the environment, transportation, and healthcare. Anti-corruption measures and procedures were undermined and unenforced (Shaub 2020). Trump even left the US Postal Service, the oldest existing federal agency, out of the massive public financing package addressing the economic effects of the pandemic. This is symbolic of the entire attitude to the utility, or from the anti-federalist perspective, the futility, of the federal government.

Crucially in the context of the COVID-19 pandemic, the federal pandemic warning system was dismantled as a leftover from Obama; the CDC, the main federal government agency charged with preparing for and managing disease outbreaks, had its budget gutted; and Trump left the states and their governors to fend for themselves without much of any real federal policy or plan to speak of (Haffajee and Mello 2020) Simultaneously, Trump also exhibited a complete disinterest in collaborating with other countries, including longtime allies, in addressing the pandemic. This would have been to resurrect the dreadful global international order that he has consistently decried. He attacked and then withdrew the United States from the World Health Organization for being too pro-China and as if it were to blame for his own months' long passivity. On the positive side, the US Army Corps of Engineers and the vaccine development program, Operation Warp Speed, provided the basis for excellent responses, respectively, in helping hospitals deal with tenting hospitalized patients and promising a conclusion to the pandemic though mass immunity by way of vaccination. Made possible only because of federal government action in both cases, the latter, however, then fell foul of the continuing lack of proper coordination between the federal government and the states in rolling out the vaccines beginning in December 2020 (e.g., Smith and Choi 2021).

Repeatedly, Trump returned again and again in 2020 to the populist idea that the pandemic was the product of travelers, particularly foreign ones. He substituted this for the idea that once the virus was present within the country, it was tracing and isolating people (as in "community spread") who test positive that should take center stage rather than simply restricting international travel. Returning to his obsession with immigrants, he obscured his mismanagement of the pandemic by announcing a total ban on immigration as if immigrants were the continuing source of infection (*NY Times* 2020). Yet as time went by the federal government seemed less rather than more effective in testing for the virus (Lim 2020). The states were left carrying the can, so to speak. Finally, in his daily press conferences early in the pandemic Trump was in full populist mode: accusing hospital staff of pilfering face masks and other Personal Protective Equipment, contradicting the public health experts, peddling his own doubtful cures like a snake-oil salesman, and instead of showing any grasp of the managerial issues facing his government, verbally assaulted the media representatives present and dispensed advice that was the opposite of that he had given the day before (e.g., Wright and Campbell 2020; Rucker and Costa 2020; Lipton et al. 2020).

Conclusion

In the end, therefore, President Trump's populist potential in addressing a national problem like the pandemic through coordination between federal and state governments proved impossible to realize. This was not simply because the populism he represents is inherently oppositional and rhetorical rather than practical. It was more because a constituency committed to limiting rather than empowering the federal government elected Trump. At the same time, and reflecting this anti-federalist electorate, he was also the prisoner of a longstanding set of ideological-institutional trends in the US that have systematically weakened the role of the federal government in managing *across* other tiers of government and thus laid the groundwork for the failures manifest in the US response to the COVID-19 pandemic of 2020. The displacement of a polyphonic version of federalism by a rigidly dualist one since the 1980s arguably undermined the possibility of an adequate federal response, notwithstanding Trump's own problematic approach to the pandemic crisis. In this context his national-populism could only ever be a fake version of the real thing. The dualist vision of US federalism that Donald Trump inherited finally proved fatal for a large number of Americans.

References

Agnew, J 1994. The Territorial Trap: The Geographical Assumptions of International Relations Theory. *Review of International Political Economy* 1 (1): 53–80. doi:10.1080/09692299408434268
_____. 2005. *Hegemony: The New Shape of Global Power*. Philadelphia: Temple University Press.
_____. 2011. Dualisme contre polyphonie dans la gouvernance territoriale contemporaine. In *Gouverner les territoires*, edited by G. Bettoni. Paris: IGPDE, Ministry of Finance.

_____. 2013. Foreword to Special Issue on *Catastrophic Geographies*. *Geographical Review* 103 (4): 455–457. doi:10.1111/j.1931-0846.2013.00012.x

_____. 2018. *Globalization and Sovereignty: Beyond the Territorial Trap*. Lanham MD: Rowman and Littlefield.

_____. 2019. Spatial Uncertainties of Contemporary Governance. *Territory, Politics, Governance* 7 (4): 435–437. doi:10.1080/21622671.2019.1663391

Agnew, J., and M. Shin. 2019. *Mapping Populism: Taking Politics to the People*. Lanham MD: Rowman and Littlefield.

Allen, J. 2009. Three Spaces of Power: Territory, Networks, Plus a Topological Twist in the Tale of Domination and Authority. *Journal of Power* 2 (2): 197–212. doi:10.1080/17540290903064267

Almond, G., and S. Verba. 1963. *The Civic Culture: Political Attitudes and Democracy in Five Nations*. Princeton: Princeton University Press.

Baker, D. R., and E. Court 2020. States Sidestep Trump to Buy Millions of Virus Tests, *Bloomberg*, August 4.

Baker, P. 2020. Trump Proceeds with post-Impeachment Purge amid Pandemic. *New York Times*, April 4.

Balkin, J. M., and R. B. Siegel eds. 2009. *The Constitution in 2020*. New York: Oxford University Press.

Bandler, J. et al. 2020. Inside the Fall of the CDC, *ProPublica*, October 15.

Barron, D. J., and G. Frug. 2006. Defensive Localism. *University of Virginia Journal of Law and Politics* 21:261–291.

Bassett, M. T., and N. Linos 2020. Trump Gave up on Fighting the Virus. Now We're Paying for His Laziness, *Washington Post*, July 14.

Beaubien, J. 2020. Americans are Dying in the Pandemic at Rates Much Higher than in Other Countries, *NPR Radio*, October 13.

Bergen, P. 2019. *Trump and His Generals: The Cost of Chaos*. New York: Penguin.

Brown, W. 2019. *In the Ruins of Neoliberalism: The Rise of Antidemocratic Politics in the West*. New York: Columbia University Press.

Bump, P. 2020. The Circumstances are Wildly Different; Trump's Response Is the Same. *Washington Post*, April 5.

Cohen, M. 2020. White House Delivers Mixed Explanations on Trump's Vow to Slow down Testing, *Politico*, June 22.

Cook, N., and D. Diamond 2020. 'A Darwinian Approach to Federalism:' States Confront New Reality under Trump, *Politico*, March 31.

Deaton, A. 2020. America's Compromised State, *Project Syndicate*, July 17.

Deudney, D. H. 1995. The Philadelphian System: Sovereignty, Arms Control, and Balance of Power in the American States-union, Circa 1787-1861. *International Organization* 49 (2): 191–228. doi:10.1017/S002081830002837X

Dikshit, R. D. 1976. *The Political Geography of Federalism*. Toronto: Macmillan.

Drezner, D. 2020. Promises Made, Promises Broken: Will It Matter? *Washington Post*, April 14.

Economist 2020. America's Backwards Coronavirus Strategy, *Economist*, July 22.

Edelman, A. 2020. Trump: Government Shouldn't Rescue States and Cities Struggling under Pandemic, *NBC News*, April 27.

Edling, M. M. 2003. *A Revolution in Favor of Government: Origins of the US Constitution and the Making of the American State*. New York: Oxford University Press.

Falconer, R. 2020. Bill Barr: "All Prosecutorial Power Is Vested in the Attorney General", *Axios*, September 17.

Frank, T. 2008. *The Wrecking Crew: How Conservatives Ruined Government, Enriched Themselves, and Beggared the Nation*. New York: Holt.

Frug, G. 2001. A Legal History of Cities. In *The Legal Geographies Reader*, edited by N. Blomley, et al. Oxford: Blackwell.

Gabriel, T., and L. Lerer 2020. Who are the Voters behind Trump's Higher Approval Rating? *New York Times*, March 31.

Ganesh, J. 2020. US Republicans Must Make Peace with the State, *Financial Times*, October 21.

Gerson, M. 2020. The Dangerous Conservative Campaign against Expertise, *Washington Post*, April 9.

Glasser, S. B. 2020. A President Unequal to the Moment, *New Yorker*, March 12.

Guay, T. 2000. Local Government and Global Politics: The Implications of Massachusetts' 'Burma Law'. *Political Science Quarterly* 115 (3): 353–376. doi:10.2307/2658123

Haffajee, R. L., and M. M. Mello 2020. Thinking Globally, Acting Locally – The US Response to COVID-19, *New England Journal of Medicine*, April 2.

Hall, B. et al. 2020. Europe's Regional Leaders Chafe at Curbs from above as Second Wave Hits, *Financial Times*, October 25.

Hamilton, A., et al. 2014 [1788]. *The Federalist Papers*. Mineola NY: Dover.

Himmelstein, D. U., and S. Woolhandler. 2016. Public Health's Falling Share of US Health Spending. *American Journal of Public Health* 106 (1): 56–57. doi:10.2105/AJPH.2015.302908

Horton, R. 2020. *The COVID-19 Catastrophe: What's Gone Wrong and How to Stop It Happening Again*. Cambridge, England: Polity.

John, A. 2020. From "Birther" to "Obamagate" to "Treason." Trump Fueled His Political Career with Unfounded Theories against Obama. Here are a Few Falsehoods, *Los Angeles Times*, June 29.

Ketcham, R. ed. 1986. *The Anti-Federalist Papers and the Constitutional Convention Debates*. New York: New American Library.

Kettl, D. F. 2020. *The Divided States of America: Why Federalism Doesn't Work*. Princeton NJ: Princeton University Press.

Khazan, O. 2020. The Most American COVID-19 Failure Yet: Contact Tracing Works Almost Everywhere Else. Why Not Here? *The Atlantic*, August 31.

Kilgore, E. 2020. Red America May Blame Blue America for Coronavirus, *New York Magazine*, March 20.

Kristof, N. 2020. America and the Virus: 'A Colossal Failure of Leadership' *New York Times*, October 22.

LA Times Editorial 2020. Pretending Not to See Coronavirus Cases Won't Make Them Go Away, *Los Angeles Times*, June 25.

Lerner, A. M. et al. 2020. Preventing the Spread of SARS-CoV-2 with Masks and Other 'Low Tech' Interventions, *Journal of the American Medical Association*, October 26.

Lim, D. 2020. Coronavirus Testing Hits Dramatic Slowdown in US, *Politico*, April 14.

Lim, E. T. 2014. *The Lovers Quarrel: The Two Foundings and American Political Development*. New York: Oxford University Press.

Lipton, E. et al. 2020. He Could Have Seen What Was Coming: Behind Trump's Failure on the Virus, *New York Times*, April 11.

Madrigal, A. C., and R. Meyer 2020. A Dire Warning from COVID-19 Test Providers, *The Atlantic*, June 30.

Mann., M. 1984. The Autonomous Power of the State: Its Origins, Mechanisms and Results. *European Journal of Sociology* 25 (2): 185–213. doi:10.1017/S0003975600004239

Markusen, A. 1994. American Federalism and Regional Policy. *International Regional Science Review* 16 (1–2): 3–15. doi:10.1177/016001769401600102

Maxwell, A., and T. Shields. 2019. *The Long Southern Strategy: How Chasing White Voters in the South Changed American Politics*. New York: Oxford University Press.

Mellow, N. 2008. *The State of Disunion: Regional Sources of Modern American Partisanship*. Baltimore: Johns Hopkins University Press.

Milhiser, I. 2021. The Enormous Advantage that the Electoral College Gives Republicans, in One Chart, *Vox*, 11 January.

Mishra, P. 2020. Flailing States: Anglo-America Loses Its Grip, *London Review of Books*, July 16.

Mortenson, J. D., and N. Bagley 2020. There's No Historical Justification for One of the Most Dangerous Ideas in American Law, *The Atlantic*, May 26.

NY Times 2020. Trump Says He Will Stop Immigration, *New York Times*, April 21.

Ollmann, B., and J. Birnbaum eds. 1990. *The United States Constitution*. New York: New York University Press.

Packer, G. 2020. Underlying Conditions: We are Living in a Failed State, *The Atlantic*, June.

Parker, A., and P. Rucker 2020. How Trump's Attempts to Win the Daily News Cycle Feed a Chaotic Coronavirus Response. *Washington Post*, April 4.

Paul, D. 2002. Re-scaling IPE: Subnational States and the Regulation of the Global Political Economy. *Review of International Political Economy* 9 (3): 465–489. doi:10.1080/09692290210150680

Peel, P. 2018. The Populist Theory of the State in Early American Political Thought. *Political Research Quarterly* 71 (1): 115–126. doi:10.1177/1065912917724004

Piller, C. 2020. Official Inaction, *Science*, October 2, 24–29.

Riley, D. 2017. American Brumaire? *New Left Review* 103:21–32.

Rodden, J. 2019. *Why Cities Lose: The Deep Roots of the Urban-Rural Political Divide.* London: Hachette.

Rucker, P., and R. Costa 2020. Commander of Confusion: Trump Sows Uncertainty and Seeks to Cast Blame in Coronavirus Crisis, *Washington Post*, April 2.

Rucker, P., and C. Leonnig. 2020. *A Very Stable Genius: Donald J. Trump's Testing of America.* New York: Penguin.

Schapiro, R. A. 2005–6. Toward a Theory of Interactive Federalism. *Iowa Law Review* 91:243–317.

_____. 2009. *Polyphonic Federalism: Toward a Protection of Fundamental Rights.* Chicago: University of Chicago Press.

Schuck, P. H. 2014. *Why Government Fails So Often: And How It Can Do Better.* Princeton: Princeton University Press.

Schwartz, M. 2020. William Barr's Long Crusade, *New York Times Magazine*, June 7: 20–25, 42–5.

Shaub Jr., W. M. 2020. Ransacking the Republic, *New York Review of Books*, July 2.

Smith, M. R., and C. Choi 2021. Vaccine Rollout Confirms Public Health Officials' Warnings *AP*, January 10.

Steinhauer, J., and A. Goodnough 2020. Contact Tracing Is Failing in Many States. Here's Why, *New York Times*, July 31.

Studemann, F. 2020. Even Germany's Localism Is Questioned in the Pandemic, *Financial Times*, October 22.

Thornhill, J. 2018. How to Fix Facebook, *Financial Times*, August 6.

Treisman, D. 2007. *The Architecture of Government: Rethinking Political Decentralization.* New York: Cambridge University Press.

Trump, M. L. 2020. *Too Much and Never Enough: How My Family Created the World's Most Dangerous Man.* New York: Simon and Schuster.

Twain, M. 1894. *Tom Sawyer Abroad.* New York: Charles L. Webster.

Van der Wusten, H. 2017. Federalism and Multilevel Governance. In *The Wiley-Blackwell Companion to Political Geography*, edited by J. Agnew, et al. Chichester UK: John Wiley.

Waldmeir, P. 2020. Trump's Support Rallies Round His Flag in the Midwest. *Financial Times*, April 7.

Wehner, P. 2020. The Party of the Aggrieved, *The Atlantic*, April 21.

Weiner, R. 2020. Trump Administration's Approach to Testing Is Chaotic and Unhelpful, States' Say, *Washington Post*, July 9.

Weiner, R., and R. Helderman 2020. States are Wrestling on Their Own with How to Expand Testing, with Little Guidance from the Trump Administration, *Washington Post*, June 10.

Wheare, K.C. 1963. *Federal Government.* Oxford: Oxford University Press.

Woodward, B. 2020. *Rage.* New York: Simon and Schuster.

Wright, T., and K. Campbell 2020. The Coronavirus Is Exposing the Limits of Populism, *The Atlantic*, March 4.

WHAT ARE THE IMPACTS OF COVID-19 ON SMALL BUSINESSES IN THE U.S.? EARLY EVIDENCE BASED ON THE LARGEST 50 MSAs

QINGFANG WANG and WEI KANG⊙

ABSTRACT. The novel coronavirus (COVID-19) has caused an unprecedented impact on both the health and economic well-being of the United States. Small businesses have been particularly affected amid the outbreak of COVID-19. Based on the first nine weeks of U.S. Small Business Pulse Survey data, this study examines the impacts of COVID-19 on small businesses across the 50 most populous metropolitan areas in the United States with an individual growth model. The results demonstrate significant disparities in impacts across regions and over time. Although the death incidents directly impacted business operations, the social, economic, demographic vulnerability, and public policies were additionally critical to our understanding of these patterns. The findings directly contribute to scholarship on regional resilience under pandemic disruption from the perspective of small businesses. More specifically, our work sheds light on the relationship between regional socioeconomic vulnerability and small-business resilience. The results provide rich implications for practices and public policymaking.

COVID-19 has caused an unprecedented impact on the health and economic well-being of the United States. Although businesses of all sizes are bracing for record losses, small businesses are in particular crisis. As of 2019, small businesses accounted for 99.9 percent of U.S. firms, employed 47.5 percent of all U.S. workers, generated 1.9 million net new jobs, and represented 287,835 exporters (Williams and others 2019). Policy makers have rapidly released a number of policies and stimulus packages at both the federal and state level to lend crucial aid to businesses. However, small businesses—arguably those whom the stimulus packages were primarily aimed at—were generally unable to access these resources (for example, the *New York Times*, April 16, 26, and 27 2020). Considering the current trajectories of the economic reopening and surging coronavirus infections, the future of economic and community recovery becomes even more onerous.

Given the context, this study seeks to examine the impacts of COVID-19 on small businesses in the largest 50 U.S. metropolitan statistical areas (MSAs) from April to June 2020. Specifically, it addresses the following questions:

- How are small businesses across different regions impacted by COVID-19?
- What regional factors are related to small business's vulnerability and resilience?

"Region" is measured at the metropolitan area level in this study. The impacts are examined in three ways: the overall impacts perceived by small-business owners and impacts on business operations, to what extent the businesses have shifted to other products or services, and the business owners' expectations of returning to the same level of business as last year.

The results demonstrate significant differences across regions and over time at the initial stage of the COVID-19 outbreak. By the end of the study period, we observed a nationwide economic reopening.[1] While the death incidents directly impacted business operations, the social, economic, demographic vulnerability, and public policies were critical to our understanding of these spatial and temporal patterns. Much of the related literature focuses on postdisaster recovery from natural disasters (floods, hurricanes/typhoons, and the like). However, viral pandemics like COVID-19 pose a different set of circumstances, necessitating a reconceptualization of what resilience means within this context. Therefore, the results directly contribute to our understanding of small-business resilience (Torres and Marshall 2015; Tibay and others 2018).

How businesses react to and cope with crises depends not only on their business factors, but on where they are, who they can get help from, what resources they can access, and how their local and regional business ecosystem works. As such, the strength of any local and regional economy and existing entrepreneurship ecosystem will greatly impact the process of small businesses coping with crises. While studies of regional resilience have been dominant by regional GDP growth and employment (Palekiene and others 2015; Crescenzi and others 2016), the current study provides a unique approach to examine regional resilience through the perspectives of small businesses in coping with a pandemic-style crisis.

Furthermore, pandemic disruptions are not evenly distributed across social, economic, and racial groups (Fairlie 2020). This study sheds light on the role of social inequalities, racial disparities, and poverty as critical sources of vulnerability and lack of resilience. Thus, this research will bridge and connect literature on disaster and resilience, inequality, and regional development. As such, its interdisciplinary approach and analytical results will provide rich implications for practices and public policymaking.

THEORETICAL BACKGROUND: SMALL BUSINESS RESILIENCE AND REGIONAL RESILIENCE

The ability of small businesses to be adaptable and flexible is paramount to their capacity to absorb and respond to a crisis. Resilient businesses are those that maintain operation during a crisis and continue to thrive afterward, returning to or exceeding predisaster levels of employment and profit (Marshall and Schrank 2014). It involves both the ability to withstand systematic discontinuities as well as the capability to adapt to new risk environments. Previous research has identified a number of attributes of a business that are most important in building resilience, including business size, industry, age of business, emergency planning, financial situation, occupancy tenure, market range, and previous disaster experience (Marshall and Schrank 2014; Haynes and others 2019). For example, there is a clear and common view that leadership and management, the core competence of staff, and effective planning are what most contribute to a business's resilience (Nowell and others 2017).

Extant research has provided ample evidence of social vulnerabilities related to race and ethnicity, gender, income and poverty, education, and social isolation, among others. Social vulnerability exacerbates not only the damages caused by natural disasters, but also people's access to social and economic resources, and therefore their overall ability to cope with and recover from disaster impacts (Van Zandt and others 2012). Indeed, disaster and resilience literature has increasingly paid attention to the socioeconomic characteristics of the households that impact their ability to cope with and recover from hazard events, such as poverty, race/ethnicity, age, gender, and household composition (Wisner and others 2003; Peacock and others 2014). This has significant implications for small businesses because the postdisaster return of businesses and households is a mutually dependent and long-term iterative process (Marshall and Schrank 2014; Xiao and Peacock 2014).

Meanwhile, scholars and practitioners in medicine and public health have devoted significant efforts to defining social determinants of health and identifying resulting inequities in health outcomes. They argue that health inequalities are rooted in societal, political, and economic conditions produced and reproduced by social structures, policies, and institutions. As such, health outcomes and disparities result not strictly from individual behaviors or genetics, but also from policies, structures, and systems that circumscribe individuals' choices, access, and knowledge (Ventres 2017; Golden and Wendel 2020).

The COVID-19 pandemic has highlighted and exacerbated the well-defined inequalities within society. As COVID-19 spreads in the United States, data shows that the pandemic has disproportionately impacted vulnerable communities, such as racial minorities, older adults, homeless populations, and low-income households who already face barriers to accessing health systems and a greater risk of suffering from chronic health conditions (Tsai and Wilson 2020; Beaunoyer and others 2020; Wenham and others 2020). For example, Karaye and Horney (2020) find that minority status, language, household composition, transportation, housing, and disability predicted COVID-19 case counts in the United States. Amram and others (2020) also find that Black, Hispanic, and poorer populations are more likely to be exposed to and impacted by COVID-19. In sum, social vulnerability highlights social, economic, and geographic characteristics that determine not only risk exposure, but also community capacity to deal with, respond to, and recover from pandemic disruptions.

At the regional level, research has linked the regional capabilities to resist recessionary shocks with a combination of social and economic factors, knowledge sharing, local leadership, and human capital; a diversified economic base; a high degree of regional specialization; regional competitiveness; a supportive system of governance with science, innovations and industry, broader and multiscale institutions, national policies, and even international networks and the global division of labor (Sedita and others 2017; Di Caro and Fratesi 2018). For example, regional economic structures, resources, capabilities, and competences influence regional resilience (Martin and others 2016). The regional industrial structure is a key factor determining regional crisis resistance. In addition, the degree of sectoral specialization may impact the overall regional sensitivity as more diversification is believed to reduce the concentration of risks (Martin and Sunley 2015; Giannakis and Bruggeman 2017). At the same time, a certain degree of specialization could be beneficial to regional growth through increased competitiveness and externalities that increase long-term regional resilience (Kitsos and Bishop 2018).

Government policy directly impacts how a region copes with disruptions and promotes renewal in the regional economy (Evenhuis 2017). For disastrous events, public policy instruments are of utmost importance for precautionary planning in order to shield the region against crisis, for mitigating the worst impacts, and for transforming and reorientating the region to recover (Kakderi and Tasopoulou 2017). In the United States, assistance from the federal government is critically important to the survival of small businesses from natural disasters. For example, Haynes and others (2019) find Small Business Administration (SBA) disaster assistance is associated with higher revenue (success) rather than survival. However, when examining the effects of federal disaster assistance programs such as SBA and Federal Emergency Management Agency (FEMA) loans on the survival and success of small family businesses impacted by Hurricane Katrina, research shows that those programs were significant to their survival rather than success.

In a most recent study on China's experiences with the COVID-19 crisis, Gong and others (2020) find that the institutional experience of dealing with previous pandemic and epidemic crises, government support schemes, and regional industrial structures are important factors for the recovery and resilience rates of Chinese regions. In the United States, a wide range of policies at the federal, state, and local level have been implemented, some related to physical distancing, including stay-at-home orders, travel restrictions, and closing nonessential businesses and some committed to crisis relief, including economic stimulus and business support (Gostin and Wiley 2020). Bartik and others (2020) have highlighted the critical role of economic and public health policy measures, such as assistance through the Coronavirus Aid, Relief, and Economic Security (CARES) Act. They found that most businesses planned to seek funding through federal assistance programs; however, many anticipated problems with accessing the program. They call for better policy design and implementation, such as streamlining the application process and clarifying the eligibility criterion and loan forgiveness rules.

Based on different strands of literature, we conceptualize businesses' coping process as a dynamic interaction and collaboration between individuals, households, businesses, and communities to adapt to external shocks to ultimately survive and thrive. We hypothesize that impacts of COVID-19 on small businesses differ significantly by region as each region has different combinations of social, economic, and institutional characteristics; and the pandemic's impact on small businesses is significantly related to the factors contributing to regional long-term economic resilience. Specifically, we examine how the impacts on small businesses are contingent on the regional factors in the following dimensions:

- Compositional factors related to the sectoral/industrial structure of regional economies
- Socioeconomic vulnerability under pandemic disruption
- National and state policies in coping with COVID-19 and economic recovery

DATA, VARIABLES, AND MODELING STRATEGY

DATA

The Small Business Pulse Survey (SBPS) is a weekly survey designed and conducted by the Census Bureau specifically for providing near real-time information about the impacts of COVID-19 on small businesses. The survey results are released at the national and state level, and for the most-populous 50 metropolitan statistical areas (MSAs).[2] SBPS Phase 1 includes nine weeks of data spanning from April 26 to June 27, 2020. For this study, we follow SBPS's definition of small businesses: nonfarm, single-location businesses with 1–499 employees and receipts of at least $1,000 as of April 2020s Census Business Register. The metropolitan data comes from the American Community Survey (ACS), Integrated Public Use Microdata Series (IPUMS) (Ruggles et al. 2020), Bureau of Economic Analysis (BEA) Regional economic accounts, and secondary data on health and government responses to COVID-19. Table 1 lists all the variables and their data sources used in this study.

MEASURING COVID-19 IMPACTS ON SMALL BUSINESSES AT THE MSA LEVEL

We adopt seven measures of COVID-19 impacts on small businesses from SBPS: "Overall," "Shifting," "Revenue," "Close," "EmpNum," "EmpHour," and "Expectation." They are longitudinal, and each contains nine waves (Week 1—Week 9).

The first measure, "Overall," represents small-business owners' subjective assessment of the overall negative impact. In the questionnaire, the following

TABLE 1—DESCRIPTIONS AND SOURCE OF VARIABLES FOR THE TOP 50 MSAS

VARIABLE	DESCRIPTION	SOURCE
Dependent variables		
Overall	9 weeks of indices of Overall Negative Impact.	SBPS Apr 26, 2020—Jun 27, 2020.
Shifting	9 weeks of % of small businesses shifting to the production of other goods or services.	SBPS Apr 26, 2020—Jun 27, 2020.
Revenue	9 weeks of % of small businesses experiencing a decrease in operating revenues/sales/receipts.	SBPS Apr 26, 2020—Jun 27, 2020.
Close	9 weeks of % of small businesses temporarily closing any of its locations for at least one day.	SBPS Apr 26, 2020—Jun 27, 2020.
EmpNum	9 weeks of % of small businesses decreasing the number of paid employees.	SBPS Apr 26, 2020—Jun 27, 2020.
EmpHour	9 weeks of % of small businesses decreasing the total number of hours worked by paid employees.	SBPS Apr 26, 2020—Jun 27, 2020.
Expectation	9 weeks of months of expected recovery.	SBPS Apr 26, 2020—Jun 27, 2020.
Independent variables		
Cross-sectional variables		
HHI	Regional industrial concentration measured by the Herfindahl–Hirschman Index (HHI).	BEA Regional economic accounts 2018.
Essential	% of workers in essential industries in 2018	2014–2018 ACS microdata (IPUMS).
EssMinority	% of minority (non-non-Hispanic White) workers in essential industries.	2014–2018 ACS microdata (IPUMS).
HSI	% of jobs labeled as "high-status" in the NAICS sectors of Information (51), FIRE (Finance, insurance, and real estate) (52–53), and Professional, scientific, and technical services (54).	BEA Regional economic accounts 2018.
VI	% of jobs in the top 3 NAICS sectors (Accommodation and Food Services (72), Arts, Entertainment, and Recreation (71), and Educational Services (61)) identified to be most vulnerable to the COVID-19 pandemic.	SBPS and BEA Regional economic accounts 2018.
SVI	Social vulnerability index (SVI) constructed from 15 socioeconomic variables by the Center for Disease Control and Prevention (CDC) (Flanagan and others 2018).	2014–2018 ACS, CDC.
SVI2	Household composition and disability—Theme 2 of Social vulnerability index (SVI).	2014–2018 ACS, CDC.
Popln	Log of total population in 2018.	2014–2018 ACS.
Party	Political party of the Governor.	
SAH	Days from the enactment of the state-wise stay-at-home order to the first day of the SBPS.	SBPS and New York Times.
Longitudinal variables amid the pandemic		

(continued)

TABLE 1—CONTINUED

VARIABLE	DESCRIPTION	SOURCE
IR	Infection rate, also referred to as "R-effective", quantifies the disease's "virality" by epidemiology models.	COVID Act Now
ICI	Infected case incidence: newly weekly infected cases for every 1000 people.	New York Times county-level data on COVID-19.
DCI	Death case incidence: newly weekly death cases for every 100,000 people.	New York Times county-level data on COVID-19.
Reopen	Binary variable tracking whether the state reopened the economy.	New York Times state reopen map.
AssiNo	9 weeks of % of small businesses that have not received any federal financial assistance since Mar 13, 2020 (Apr 26, 2020—Jun 27, 2020).	SBPS Apr 26, 2020—Jun 27, 2020.

question is asked: "Overall, how has this business been affected by the COVID-19 pandemic?" The index of the study was constructed thusly: we assign score 1, 0.5, 0, −0.5, and −1 to five responses ranging from "large negative effect," "moderate negative effect," "little or no effect," "moderate positive effect," and "large positive effect," respectively; we then pool the total weekly scores for all MSAs and normalize them to be within the range of −1 to 1 so that larger scores represent larger negative impacts and smaller scores represent smaller negative impacts. As shown in the choropleth maps with quintile classification in Figure 1, this measure displays high spatial heterogeneity, with coastal MSAs generally reporting larger negative overall impacts. While some MSAs maintained their relative position (for example, Los Angeles-Long Beach-Anaheim, California, and New York-Newark-Jersey City, New York-New Jersey-Pennsylvania, fell within the fifth quintile) during these nine weeks, others experienced drastic changes from Week 1 (Figure 1(a)) to Week 9 (Figure 1(b)) (for example, New Orleans-Metairie, Louisiana, transitioned from quintile 1 to quintile 5), suggesting regional discrepancies in the experience and response to the COVID-19 pandemic.

The second measure, "Shifting," is asked in the questionnaire as follows: "In the last week, did this business shift to the production of other goods or services?" In our study, the index is constructed as the proportion of small businesses that shifted to the production of other goods or services within each MSA in the last week. We visualize its spatial patterns at Weeks 1 and 9 in Figure 2(a,b). Similar to "Overall" impacts, we observe spatial disparity in both periods: the shifting proportion ranged from 3 percent (Buffalo-Cheektowaga-Niagara Falls, New York) to 13 percent (Jacksonville, Florida) at Week 1 and 1 percent (New Orleans-Metairie, Louisiana) to 11 percent (Memphis, Tennesse-Mississippi-Arkansas) at Week 9.

a

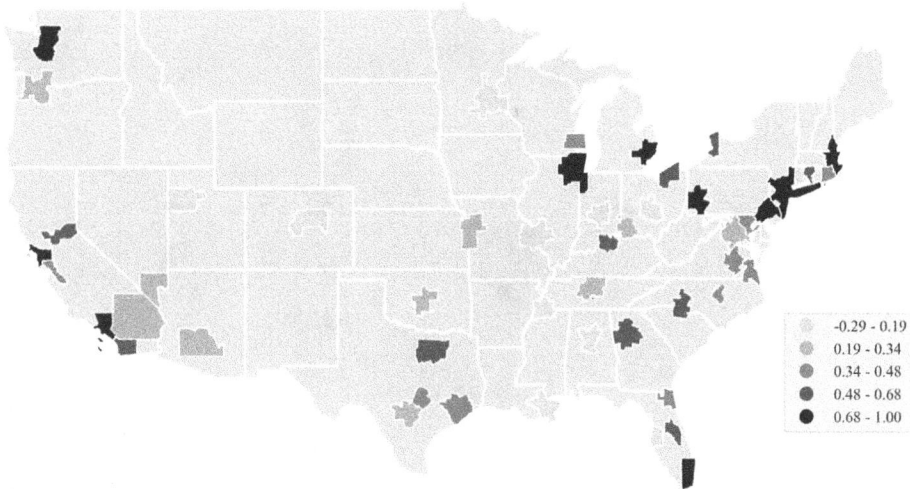

b

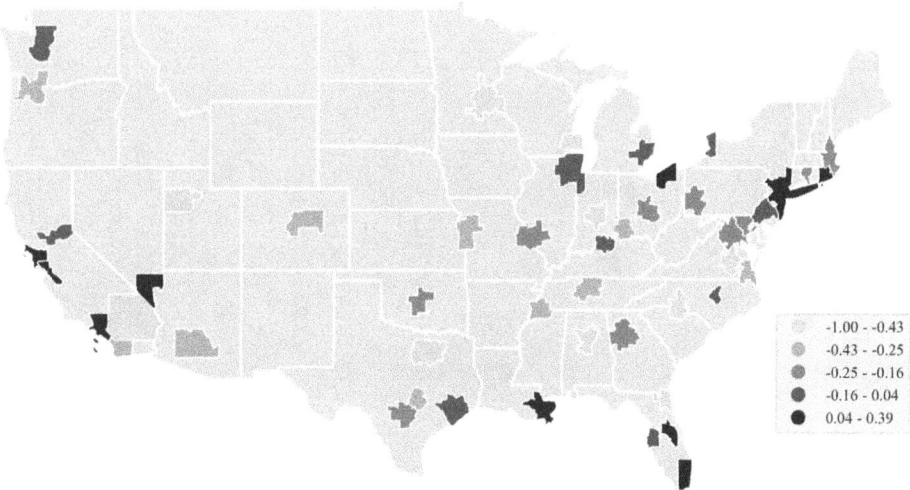

Fig. 1—Overall assessments of negative impacts on small business across top 50 MSAs: (a) Week 1: 2020/04/26-2020/05/02, (b) Week 9: 2020/06/21-2020/06/27.

The next four measures—"Revenue," "Close," "EmpNum," and "EmpHour"—evaluate impacts of the pandemic on the operation of small businesses in four aspects -(note that these measures are assessed by decreases or declines in revenue, employment size, and working hours). The questionnaire asks: "In the last week, did this business experience a change in operating revenues/sales/receipts, not including any financial assistance or loans?"; "In the last week, did this business temporarily close any of its locations for at least one day?"; "In the last week, did this business have a change in the number

a

b

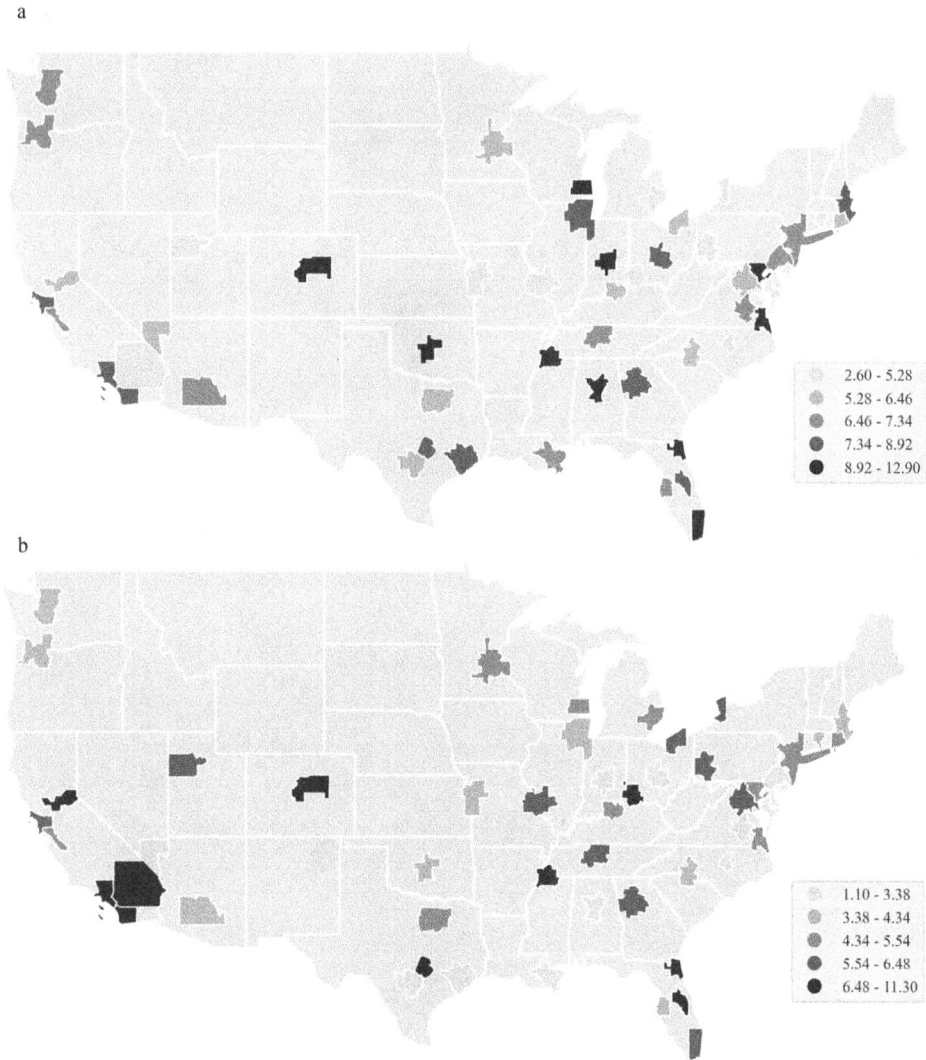

FIG. 2—Proportions of small business shifting to the production of other goods or services in the last week across top 50 MSAs: (a) Week 1: 2020/04/26-2020/05/02, (b) Week 9: 2020/06/21-2020/06/27.

of paid employees?"; and "In the last week, did this business have a change in the total number of hours worked by paid employees?".

As stay-at-home orders and the social-distancing policies were enacted in most U.S. regions before Week 1, by Week 1, the operation of many businesses was expected to be disrupted to some extent in terms of decreased revenue, temporary closure, increased layoffs, or increased furloughs. By the end of the Phase 1 SBPS (Week 9), every U.S. state had at least partially reopened its economy, and as a result, we expect to observe a decrease in the negative operational impacts by that week.

At Week 1, more than 58 percent of small businesses in each of the top 50 MSAs suffered from a decrease in revenue. By Week 9, the MSA most suffering was Orlando-Kissimmee-Sanford, Florida, where 55.7 percent of small businesses reported a revenue decrease in the past month. While the negative impact on revenue in every MSA abated during the survey period, the extent varied across space. For instance, compared with New York-Newark-Jersey City, New York-New Jersey-Pennsylvania, which experienced a 38.9 percent decrease (from 79.4 percent to 40.5 percent), Los Angeles-Long Beach-Anaheim, California, which also experienced high infection rates at the start of the pandemic, only obtained a 27.3 percent decrease (from 78.3 percent to 51 percent). Therefore, at Week 9, the Los Angeles MSA still fell within the fifth quintile, leading the other MSAs on this measurement. We observe a similar pattern for the other three variables assessing operational impacts. Week 1 demonstrated a high spatial heterogeneity. Although every MSA has seen a declining trend in each dimension, which means a better situation in their performances, the extent varied.

The seventh index, "Expectation," measures the length of time in months thatsmall-business owners expected before recovering to the same business level as last year. In the questionnaire, the following question is asked: "In your opinion, how much time do you think will pass before this business returns to its normal level of operations relative to one year ago?" We construct this index by first assigning scores (0.5, 2.5, 5, 10, 20, 0) to six possible responses of "1 month or less," "2–3 months," "4–6 months," "more than 6 months," "not returning to normal," and "little or no effect"; we then score for each MSA per week. As shown in Figure 3, generally, the coastal MSAs were more pessimistic about the recovery, although the growth for some MSAs was slower compared with others. For instance, San Diego-Carlsbad, California, expected a 6.95-month and 7.48-month recovery at Week 1 and Week 9, respectively.

We visualize the temporal trend of the average MSA-level impacts for each of the seven dimensions in Figure 4. The overall assessment of negative impacts, the proportion of small businesses shifting to the production of other goods or services, as well as the negative operational impacts generally followed a declining trend. In contrast, small businesses expected a longer recovery period. The changes were most drastic in the first several weeks.

The temporal trends are further confirmed by the changing statistical distributions at Week 1 (blue), 5 (orange), and 9 (green) in the diagonal plots in Figure 5. The nondiagonal plots of Figure 5 visualize the pairwise relationships between the seven impact measures. We also note that "Overall" negative impacts and "Expectation" to recover exhibit a steady, modestly positive relationship, with the weekly correlation coefficient within the range of 0.499 to 0.719. It indicates that the MSAs where small businesses identified themselves as being strongly negatively impacted by the COVID-19

a

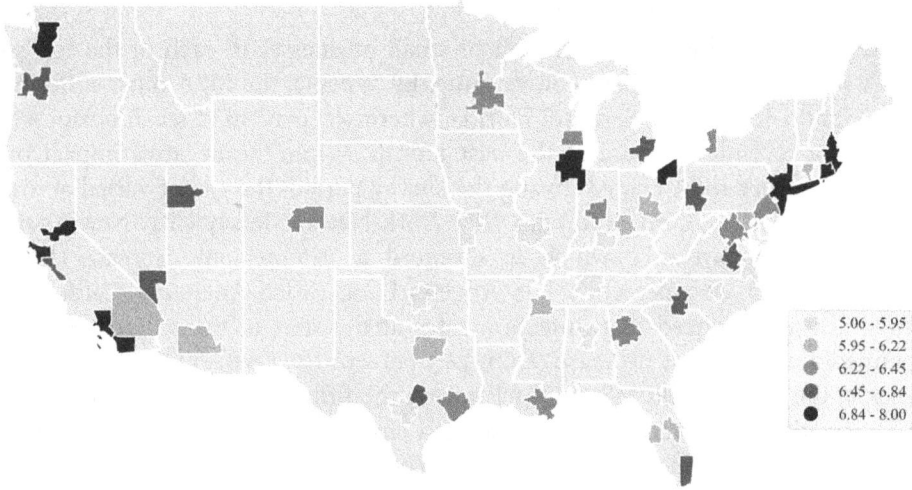

b

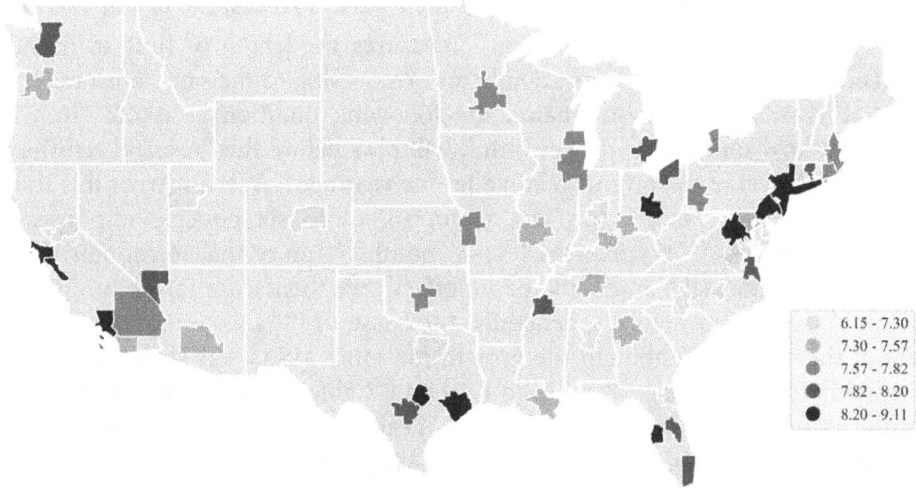

FIG. 3—The number of months expected by small business owners to recover to the same level in the last year: (a) Week 1: 2020/04/26-2020/05/02, (b) Week 9: 2020/06/21-2020/06/27.

tended to be those where small businesses were more worried about a lengthy recovery.

INDEPENDENT VARIABLES: THE REGIONAL FACTORS

Economic structure

The first index is the Herfindahl–Hirschman Index (HHI), a widely adopted measure for regional industrial concentration (Mack and others 2007).[3] The HHI

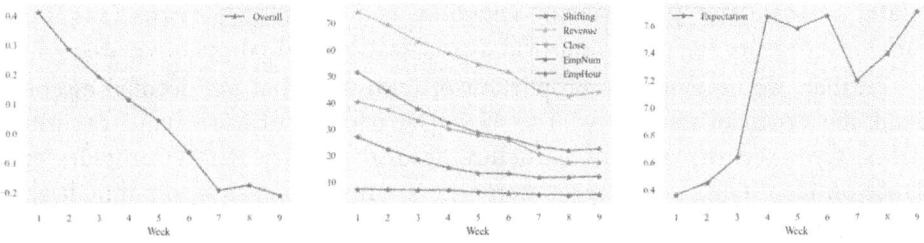

FIG. 4—Average trends of COVID-19 impacts on small businesses across the top 50 MSAs.

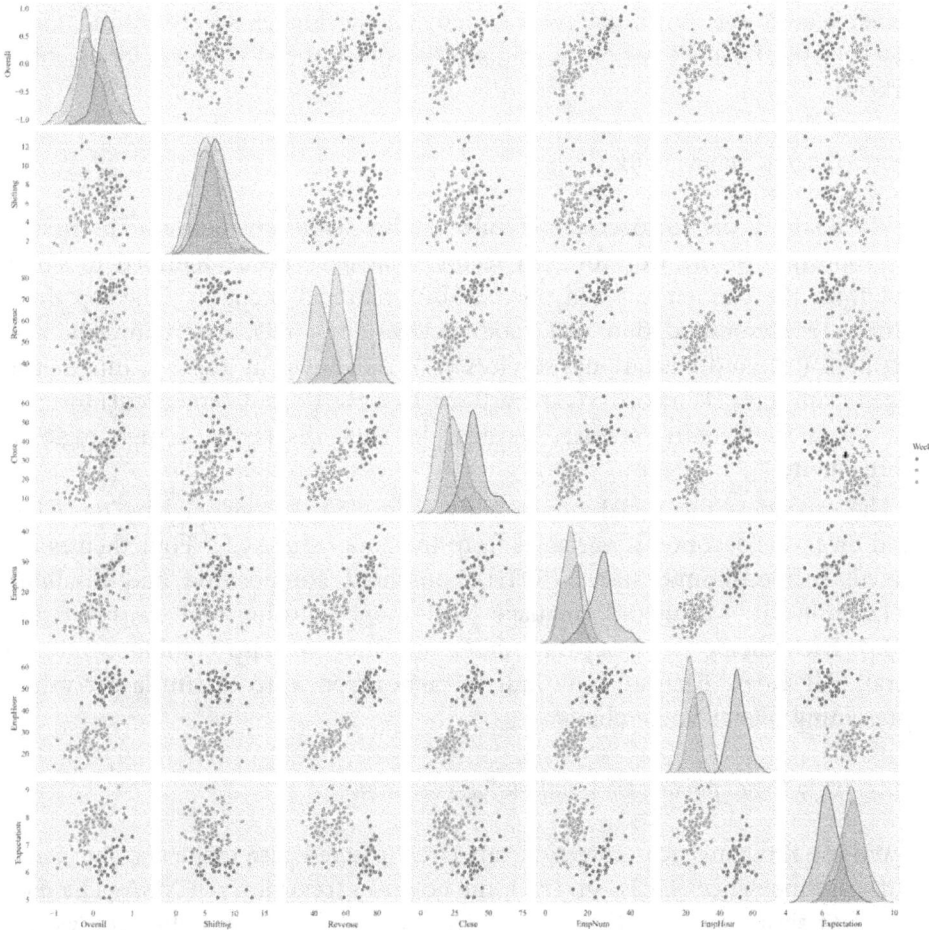

FIG. 5—Statistical distributions of and pairwise relationships between the seven indices of impacts at Weeks 1, 5, and 9.

for the 50 most populous MSAs in 2018 ranged from 0.067 to 0.1. The second variable is the proportion of "high-status" industries considered to be high-wage and high-skilled, including information (51), FIRE (finance, insurance, and real estate) (52–53), and professional, scientific, and technical services (54). Both variables measure regional industrial diversification and competitiveness.

Further, we measure the proportion of industries that are deemed essential amid the COVID-19 pandemic. The U.S. Department of Homeland Security's (DHS) Cybersecurity and Infrastructure Security Agency (CISA) provides guidance on identifying essential workers who are deemed critical to public health, safety, and economic and national security. We further look at the racial composition of essential workers in each MSA. Out of all essential workers in an average MSA, minority workers made up 41.33 percent in 2018. However, it had a rather wide range—13.90 percent to 72.19 percent—with MSAs in the Sunbelt region and northeast coast occupying the higher spectrum and those located in the Rustbelt region having smaller proportions of minority essential workers.

Socioeconomic vulnerability

We quantify the socioeconomic vulnerability with two metrics. The first is the proportion of the workforce in sectors that have been identified to suffer most from the pandemic based on the SBPS national accounts. The top three sectors are accommodation and food services (72); rrts, entertainment, and recreation (71); and educational services (61). As shown in Table 2, this metric ranged from 10.63 (Hartford-West Hartford-East Hartford, Connecticut) to 24.74 (Las Vegas-Henderson-Paradise, Nevada) in 2018, displaying a high regional heterogeneity.

The second is the social-vulnerability index (SVI) developed by the CDC based on 15 socioeconomic variables from the 2014–2018 ACS.[4] Four themes are formed: socioeconomic status (SVI1), household composition and disability (SVI2), minority status and language (SVI3), and housing and transportation (SVI4). Each theme has a separate index and they are also combined into an overall SVI index. Each of these indices ranges from 0 to 1, with larger values representing higher vulnerability.

COVID-19 trend, public policy and business assistance

We use three metrics—infection rate (IR), infected case incidence (ICI), and death-case incidence (DCI)—to track the COVID-19 trend in each MSA. The data (Figure 6) shows that the IR has been steadily increasing over the nine weeks, indicating a trend of growing virality. It started with a below 1 value at Week 1, 0.96, indicating each person with COVID-19 would transmit the virus to an average of 0.96 other people, according to this dataset. This below 1 value is a good indicator of shrinking infection rates of COVID-19. However, starting at Week 6, the average IR surpassed 1 and continued to increase, suggesting rapid infection rates during this period of the pandemic. Comparatively, the ICI, which

TABLE 2—RESULTS OF INDIVIDUAL GROWTH MODELS

	PARAMETER	OVERALL	SHIFTING	REVENUE	CLOSE	EMPNUM	EMPHOUR	EXPECTATION
Initial status, π_{01}	Intercept	0.418^{**}	7.297^{**}	71.9^{**}	40.98^{**}	21.75^{**}	44.86^{**}	6.91^{**}
		(0.006)	(0.285)	(0.73)	(1.09)	(0.705)	(0.700)	(0.106)
	HHI β_{01}	-2.353	5.228	-135.8	54.16	47.28	-60.78	-0.251
		(5.196)	(22.11)	(68.6)	(133.8)	(75.64)	(61.37)	(11.54)
	Essential β_{02}	-0.003	-0.004	-0.175	-0.13	-0.132	-0.118	-0.014
		(0.014)	(0.058)	(0.181)	(0.358)	(0.201)	(0.161)	(0.031)
	EssMinority β_{03}	0.0007	0.0095	0.006	0.071	0.004	0.092^{*}	0.007
		(0.0029)	(0.012)	(0.039)	(0.075)	(0.043)	(0.034)	(0.006)
	HSI β_{04}	0.007	-0.010	-0.189	0.479	-0.021	-0.076	0.004
		(0.012)	(0.054)	(0.164)	(0.31)	(0.178)	(0.147)	(0.027)
	VI β_{05}	0.029^{*}	-0.050	0.394^{*}	0.44	0.145	0.265	0.041
		(0.012)	(0.049)	(0.153)	(0.304)	(0.17)	(0.136)	(0.026)
	SVI β_{06}	0.139	2.261	6.867	6.396	3.082	-3.260	0.589
		(0.367)	(1.512)	(4.678)	(9.20)	(5.167)	(4.183)	(0.789)
	SVI2 β_{07}	-0.236	-4.46^{**}	-5.293	-9.49	-1.195	3.700	-1.67^{*}
		(0.339)	(1.416)	(4.419)	(8.76)	(4.903)	(3.941)	(0.749)
	Popln β_{08}	0.135^{*}	0.325	0.538	1.189	-0.012	-0.367	0.184
		(0.051)	(0.215)	(0.669)	(1.302)	(0.737)	(0.598)	(0.112)
	Party (Republican) β_{09}	-0.076	0.436	-1.333	-3.033	-1.316	-1.678^{*}	-0.269
		(0.061)	(0.260)	(0.803)	(1.567)	(0.883)	(0.718)	(0.135)
	SAH β_{10}	-0.002	0.0009	0.012	0.146	0.097	0.037	0.002
		(0.004)	(0.015)	(0.047)	(0.093)	(0.052)	(0.042)	(0.008)
Rate of change, π_{11}	Intercept β_{10}	-0.08^{**}	-0.36^{**}	-3.43^{**}	-2.69^{**}	-0.887^{**}	-2.593^{**}	0.097^{**}
		(0.006)	(0.068)	(0.158)	(0.141)	(0.127)	(0.156)	(0.019)
Time-varying variable	IR β_{20}	0.0099	-0.335	-1.67	-1.92	-0.353	1.495	-0.545^{*}
		(0.087)	(0.854)	(2.03)	(1.93)	(1.7)	(2.004)	(0.252)

(continued)

TABLE 2—CONTINUED

	PARAMETER	OVERALL	SHIFTING	REVENUE	CLOSE	EMPNUM	EMPHOUR	EXPECTATION
ICI	β_{30}	0.004	−0.007	0.07	0.083	0.072	0.045	0.015
		(0.004)	(0.031)	(0.084)	(0.092)	(0.076)	(0.079)	(0.011)
DCI	β_{40}	−0.0007	−0.103	0.188	0.264	0.302*	0.38*	−0.008
		(0.008)	(0.061)	(0.16)	(0.175)	(0.145)	(0.153)	(0.022)
Reopen	β_{50}	−0.025	−0.122	−3.04**	−1.95**	−2.44**	−2.68**	0.013
		(0.033)	(0.332)	(0.787)	(0.704)	(0.608)	(0.755)	(0.095)
AssiNo	β_{60}	−0.0009	−0.018	0.077**	0.017	0.184**	0.230**	−0.022**
		(0.001)	(0.012)	(0.028)	(0.025)	(0.022)	(0.027)	(0.003)
Variance Components and Explained Variance								
Level 1 Within-MSA	$\sigma_{\varepsilon}2$	0.038**	3.817**	21.01**	15.81**	11.65**	18.85**	0.294**
		(0.003)	(0.287)	(1.496)	(1.126)	(0.916)	(1.1445)	(0.021)
	R12	0.56	0.17	0.85	0.82	0.71	0.84	0.42
Level 2 Between-MSA	$\sigma 02$	0.026**	7.7e-8**	2.55**	17.76**	4.37**	1.27**	0.109**
		(0.007)	(0.137)	(1.15)	(4.461)	(1.42)	(0.983)	(0.033)
	R22	0.43	0.99	0.5	0.54	0.24	0.65	0.60
Goodness-of-fit (stochastic parts)								
	AIC	33.80	1941.7	2711.2	2652.1	2464.6	2644.4	890.3
	BIC	115.22	2023.1	2788.5	2729.4	2546	2725.8	967.6

*p < .05; **p < .01; inside the parentheses are standard errors.

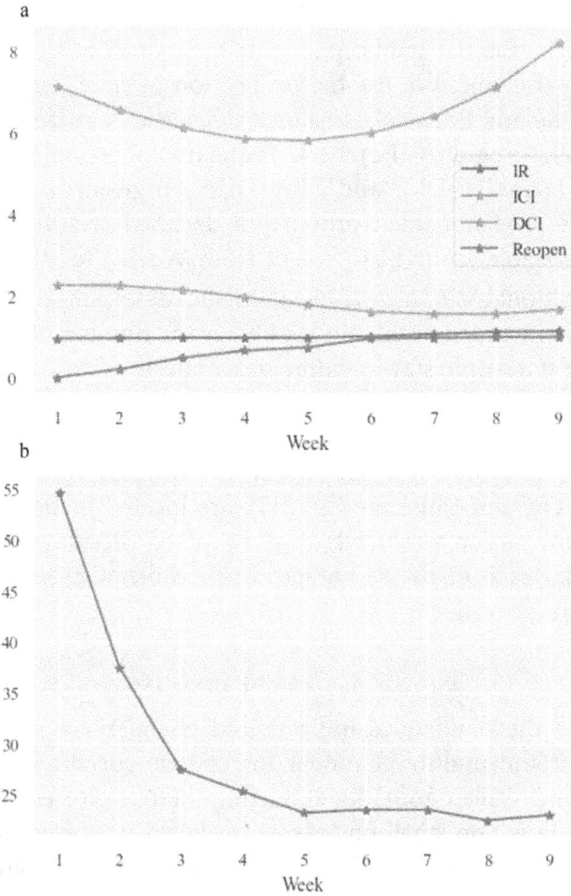

FIG. 6—Average trend of (a) COVID-19 severity, reopen policy, and (b) proportion of small businesses federal financial assistance across the top 50 MSAs.

measures the newly weekly infected cases for every 1000 people, experienced a decline from Week 1 to Week 5 (05/24/2020–05/30/2020), then an increase from Week 5 to Week 9. The DCI, which measures newly weekly death cases for every 100,000 people, steadily decreased from Week 1 to Week 7 (06/07/2020–06/13/2020), and then rebounded to 1.67 at Week 9.

As reopening the economy is argued to be an important policy to help small businesses to recover from the pandemic, we track the reopening status in each U.S. state and incorporate this change in our model to investigate whether it is correlated with the measured impacts on small businesses. As shown in Figure 6, at Week 1 (04/26/2020–05/02/2020), none of the states in which the top 50 MSAs are located has reopened the economy. Starting from Week 2, 20 percent of MSAs reopened the economy. The number continued to grow to 100 percent at Weeks 8 and 9.

We also track the status of federal financial assistance for each MSA as this is arguably the most essential public policy to help small businesses stay solvent and survive the pandemic. We use the proportion of small businesses that have not received any federal financial assistance since March 13, 2020. The assistance covers three main programs: Paycheck Protection Program (PPP), Economic Injury Disaster Loans (EIDL), and SBA Loan Forgiveness, along with other federal programs. The nonaided proportion declined steadily in the first five weeks but stayed stable around 23 percent from Week 5 to Week 9.

Two cross-sectional variables relevant to the state public policies amid the pandemic are also included in the model. One is the number of days between the enactment of the state-wide stay-at-home order and the first day of the SBPS; the other is the political party of the state governor. The former metric is largest (38 days) for the six MSAs in California, as California was the first state to execute a state-wide stay-at-home order on March 19, 2020. MSAs like Oklahoma City, OK and Salt Lake City, UT are located in states where a state-wide stay-at-home order was never put in place. Besides, it might not be a coincidence that each of the seven states that did not execute such an order have a Republican governor.

INDIVIDUAL GROWTH MODELING

To examine how the pandemics impacts and changes are related to regional factors, we adopt the multilevel model for change (or the individual growth model) (Singer and Willett 2003) for modeling the trajectories of the impacts of the COVID-19 pandemic on small businesses in the 50 most-populous MSAs. The individual growth model is formally defined as a two-level hierarchical model as shown in Equation (1), where i represents MSA and t represents time. Level 1 models the individual growth trajectory of each MSA as a function of time ($Week_{ti}$) and a series of longitudinal variables ($X2_{TI}, ..., XK_{ti}$; *longitudinal variables amid the pandemic* in Table 1). At Level 2, differences among MSAs are modeled with a set of contextual variables ($Z1_1, ..., ZM_i$; *cross-sectional variables* in Table 1). We model the intercept (π_{0I}) as random effects, while treating the rate of change (π_{1I}) of the individual growth trajectory and the coefficients on longitudinal variables ($\pi_{2I}, ..., \pi_{Ki}$) as fixed effects. We also attempt to incorporate the potential temporally correlated error structure in the model. The error is modeled as a first-order autoregressive process. Each model is fitted using the restricted maximum likelihood (REML) estimation technique.

$$\text{Level 1}: \quad y_{ti} = \pi_{0I} + \pi_{1I}Week_{ti} + \pi_{2I}X2_{TI} + \ldots + \pi_{Ki}XK_{ti} + \varepsilon_{t.} \tag{1}$$

$$\text{Level 2}: \quad \pi_{0I} = \beta_{00} + \beta_{0I}Z1_1 + \ldots + \beta_{0M}ZM_i + \zeta_{0I}$$

$$\pi_{1i} = \beta_{10}$$
$$\pi_{2i} = \beta_{20}$$

$$\dots$$

$$\pi_{Ki} = \beta_{K0}$$
$$\varepsilon_{ti} \sim N(0, \sigma_\varepsilon^2) \text{ and } \zeta_{0i} \sim N(0, \sigma_0^2)$$

We center the continuous variables as a potentially effective way to avoid high collinearity (Finch and others 2019). Specifically, for each of the longitudinal predictor variables "AssiNo," "IR," "ICI," and "DCI," we apply grand mean centering, where the overall mean is subtracted from all values of a variable. We also rescale the time predictor $Week_{ti}$ so that it starts at 0 instead of 1. Combining these two rescaling efforts, we would be able to interpret the intercept π_{0i} as the baseline or initial status of the response variable. We also resort to the variance inflation factor (VIF) to assess the degree of multicollinearity and make sure it is smaller than 5 to avoid potential multicollinearity issues (Fox 2015).

REGRESSION RESULTS

"OVERALL" NEGATIVE IMPACTS

As shown in the "Overall" column of Table 3, the average trajectory of the "Overall" perceived negative impacts started with 0.418 at Week 1 and decreased over time by 0.08 per week. At the initial stage, businesses tended to report higher overall negative impacts in the most-populous MSAs. Larger cities like New York, Los Angeles, and San Francisco likely experienced early viral shocks due to higher rates of international travel. In addition, for each MSA, its initial status is related to the percentage of jobs in vulnerable industries (VI) and population size. Specifically, a 1 percent increase in the proportion of jobs in these industries is associated with a 0.029 increment of an initial overall assessment of negative impacts. Figure 7 depicts the predicted value of overall negative impacts in three prototypical trajectories for given VI levels from Week 1 to Week 9. As time progresses, the negative impacts are perceived to decrease; however, higher VI predicts higher negative impacts.

None of the time-varying predictors on COVID-19 evolution and public policies are significant in the full model. This indicates that business owners' assessment of the overall negative impact was unrelated to how the COVID-19 pandemic unfolded in each region. The federal financial assistance and the reopening of the economy were also likely to be irrelevant factors.

"SHIFTING" TO THE PRODUCTION OF OTHER GOODS OR SERVICES

The average trajectory of "Shifting" started with 7.297 percent in Week 1 and decreased by 0.36 percent every week. The only significant variable is the level-2

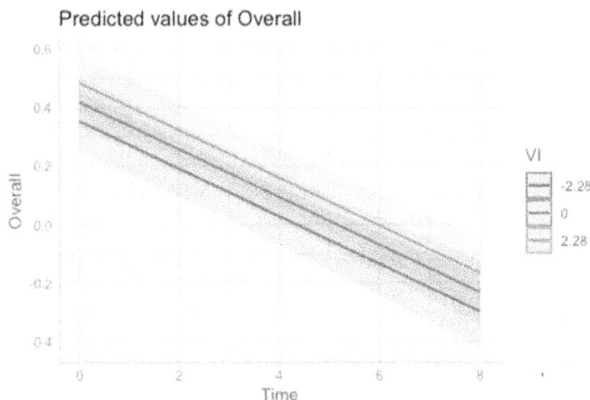

FIG. 7—Predicted overall negative impacts changes with VI from Week 1 to Week 9.

variable "SVI2" (in addition to the time variable). A larger value of "SVI2" represents a higher percentage of nonworking-age residents, residents with a disability, and single-parent households in a MSA. Since its coefficient is negative, MSAs with higher social-demographic vulnerability tend to have a lower initial "Shifting" rate. Figure 8 depicts how the predicted capability of "Shifting" production changes with SVI2 over time.

EXPERIENCING "REVENUE" DECREASE

The average trajectory of "Revenue" decrease saw a declining trend with an initial level of 71.9 percent and a decline rate of 3.43 percent per week. Similar to "Overall" negative impacts, for an average MSA, the initial proportion of small businesses experiencing a decrease in operating "Revenues" is positively related

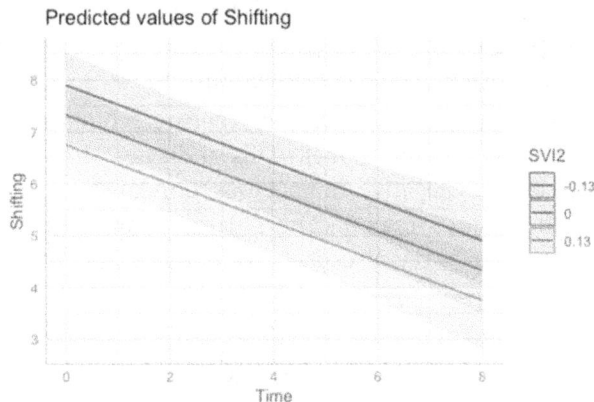

FIG. 8—Predicted shifting production with SVI2 from Week 1 to Week 9.

to the percentage of vulnerable industries (VI). A 1 percent increase is associated with a 0.394 percent reduction in "Revenue" at Week 1.

Furthermore, an average MSA that reopened the economy had a smaller "Revenue" reduction (by 3.04 percent) compared with nonopening MSAs in the same week. The other variable, "AssiNo," represents the percentage of small businesses that have not received any federal financial assistance since March 13, 2020. Unsurprisingly, more assistance is associated with less "Revenue" reduction. Specifically, a 1 percent increase in small businesses without assistance is associated with a 0.077 percent increment in small businesses experiencing a "Revenue" decrease. The relationship reflects the significant role of federal financial assistance to small businesses.

"CLOSE"

"Close" measures the proportion of small businesses temporarily closing any of their locations for at least one day in a given MSA. For the surveyed 9 weeks, the average MSA trajectory started at 40.98 percent and decreased at a rate of 2.69 percent every week. Unsurprisingly, reopening the economy had a positive effect on preventing small businesses from temporarily closing throughout all of the surveyed weeks. Specifically, an average MSA would have 1.95 percent fewer small businesses temporarily closing any of its locations for at least one day if it reopened the economy. Over time, as each State opened its economy, the proportion of closing decreases significantly.

REDUCTION OF PAID EMPLOYEES

"EmpNum" represents the percentage of small businesses reducing the number of paid employees. At Week 1, 21.75 percent of small businesses had reduced paid employees. This percentage declined at a rate of 0.887 percent per week. Reopening the economy and providing financial assistance to small businesses are significantly associated with a decrease in employment reduction. In addition, for each week, an increase of 1 in the new weekly death cases for every 100,000 people was associated with an increase of 0.302 percent small businesses suffering the loss of paid employees.

REDUCTION OF WORKING HOURS

"EmpHour" represents the percentage of small businesses decreasing the total number of working hours by paid employees. An average trajectory of "EmpHour" began at 44.86 percent in Week 1 and declined at the rate of 2.6 percent per week. Similar to "EmpNum," reopening the economy and providing financial assistance is positively related to the reduction of total working hours. At the same time, an increase of 1 in new weekly death cases for every 100,000 people was associated with a 0.38 percent increment of "EmpHour" at any week. Furthermore, the initial status of "EmpHour" was positively related to "EssMinority," the percentage of minority workers in essential industries. In other words, at the beginning of the survey period, MSAs with more minority workers in essential industries suffered more in terms of a higher percentage of small businesses decreasing the total number of hours worked by paid employees. In the following weeks, because the estimated coefficient on the rate of change is significant and negative, the "EmpHour" trajectory of each MSA declined at the rate of 2.6 percent per week irrespective of the differing initial statuses. That is, while all MSAs gradually recovered at the same rate, the MSAs with more minority workers that suffered more initially, also suffered more than other MSAs in the following weeks. In addition, MSAs located in a state with a Republican governor had a lower level of working hour reduction.

"EXPECTED" RECOVERY

"Expectation" indicates the number of months that a small-business owner expected until their business returned to prepandemic levels. The average value started at about 6.91 months in Week 1. As time went on, the expected months grew at an average of 0.097 per week. The perceived faster recovery is associated with a lower social-demographic vulnerability (SVI2). The expected recovery is also significantly associated with the changing infection rates and federal financial assistance to small businesses: the infection rate is negatively associated with the expected recovery at each week and MSAs with more small businesses receiving federal financial assistance were less optimistic about the recovery.

DISCUSSION

First, the industrial structure has minimal impacts on businesses if measured at the regional level. Previous studies argue that the sectorial structure of the regional economy, such as the industrial specialization and diversification, is a key factor determining regional crisis resistance (Martin and Sunley 2015; Kitsos and Bishop 2018). Moreover, a higher level of regional competitiveness, for example, a higher proportion of high-skilled labor force and knowledge- and technology-led sectors, can mitigate external disturbance (Sedita and others 2017; Di Caro and Fratesi 2018). In our study, however, none of the industrial structural factors—the degree of regional economic diversification, proportion of high-status industries, and the essential industries—is significant. Nonetheless, we want to caution that existing studies are focused on long-term regional economic growth. Since this study is based on the data at the early stage of COVID-19, it is warranted in the future to examine the impacts over a longer time span.

Second, a region's economic, social, and demographic vulnerability is significantly associated with small businesses' perception of recovery and overall impacts of COVID-19, as well as their decrease in revenue and working hours.

COVID-19 disruptions do not affect all businesses equally (Gong and others 2020). For example, in their survey of 5,800 small businesses between March 28 and April 4, 2020, in the United States, Bartik and others (2020) found that firms in the most exposed industries, such as restaurants, tourism, and personal services, face extreme difficulties to staying in business. Supporting similar patterns, we find a MSA's higher proportion of VI is significantly associated with higher negative overall impacts and revenue reduction (Figure 9).

Extant literature has shown that racial minorities, the poor, and the old suffer more from COVID-19 due to their socioeconomic vulnerability (see Amram and others 2020). In this study, the overall social-vulnerability index (Flanagan and others 2018) is not significant. However, a higher percentage of nonworking-age residents, residents with a disability, and single-parent households in a MSA (SVI2) is associated with a lower probability of business's shifting to other products/services and negative perception of rapid recovery. Figure 10 shows

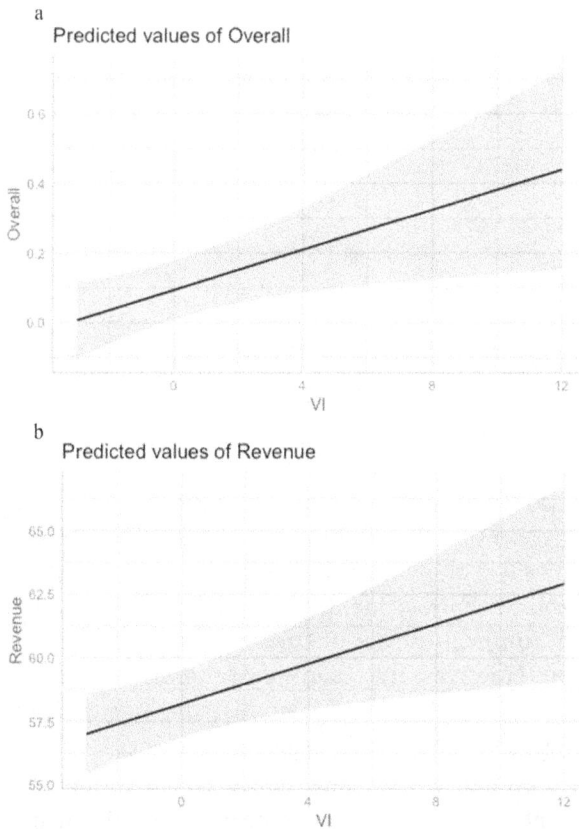

FIG. 9—Predicted values of (a) overall negative impacts and (b) revenue reduction with the change of the VI proportion in an MSA.

a
Predicted values of Shifting

b
Predicted values of Expectation

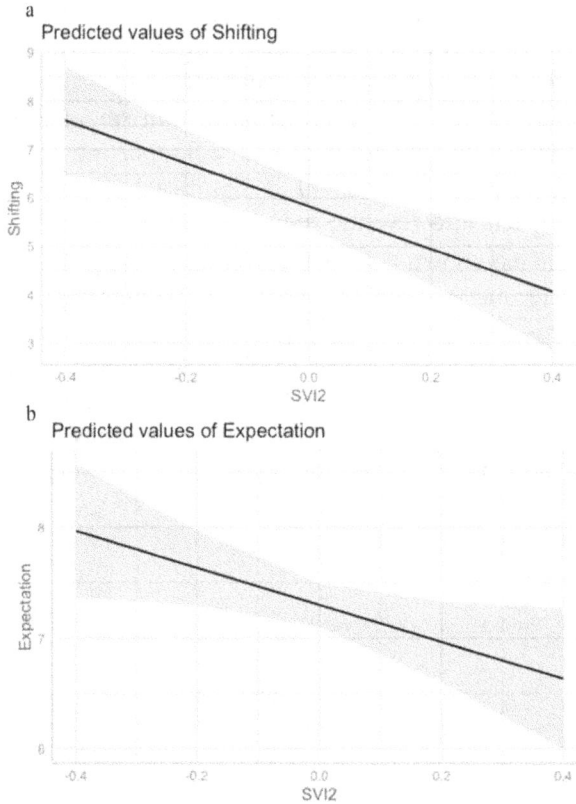

FIG. 10—Predicted values of (a) shifting production and (b) expectation of recovery (b) with the change of the SVI2 in an MSA.

the predicted values of (a) shifting production and (b) expectation of recovery with the increase of the SVI2 in an MSA.

The impacts on business shifting production are alarming. Under COVID-19 and the lockdown policies, the crisis not only leaves many businesses struggling for survival, but also forces some to look for alternative strategic paths. As such, the pandemic has generated innovations, encouraging businesses to identify new business models that will allow them to survive through the crisis, particularly to look for digital replacements or identify ways of safely delivering their products and service with minimal physical contact (Donthu and Gustafsson 2020; Seetharaman 2020). However, the socio-demographic vulnerabilities in a region present a significant barrier for small businesses to shift their productions. This has significant long-term impacts on regional resilience.

Furthermore, although we did not find a relationship between the regional proportion of essential businesses and impacts on small businesses, we indeed find a higher percentage of racial minorities in the essential industries is significantly related to a decrease in businesses' working hours by paid employees.

Out of all essential workers in an average MSA, minority workers made up 41.33 percent, with a rather wide range: 13.90 percent to 72.19 percent. Considering the concentration of minority labor force in the essential businesses, the impacts could be significantly different across regions with different proportions of the essential industries.

Third, the health situation in a region is closely related to business operational variables. While the general infected case incidence (newly weekly infected cases for every 1000 people) seems unimportant, the death-case incidence is associated with reduced business operations, and the speed of infection is related to the pessimistic perception of the future. Like the physical interruption caused by natural disasters, the reduction of employment and working hours reflects the direct impacts of pandemic disturbance of business operations.

Fourth, public policy factors are not related to the perception of overall impacts; however, they are significantly related to business operations. Bartik and others (2020) argue that the damage to our economy and its network of small businesses caused by COVID-19 will remain large if the crisis lasts longer. They suggest that reopening the economy safely at a quicker pace to shorten the economic shutdown is largely beneficial. Our study finds that reopening the economy directly increases revenues or sales, the size of employment, and working hours for small businesses, as well as reduced business shutdown. Figure 11 depicts the predicted value of revenue reduction, business shutdown, and reduction of employment with a reopening policy in place from Week 1 to Week 9.

Research has long argued that public policy in regional resilience and federal assistance programs have been essential in helping small businesses to survive or recover (Torres and others 2019). This study finds as more businesses in a region received federal assistance, the region experienced less business shutdown, an increase in employment and working hours, but an expectation of prolonged time to recover. The relationship between government assistance and impacts on business operation indicates the benefits gained by the businesses and region from federal business-aid programs. Figure 12 depicts the predicted value of revenue reduction and decrease of employment with the increase of nonaided businesses in a MSA.

Clearly, there has been tension between disease-containment policies and the speed of economic recovery as part of the resilience at different levels (Gong and others 2020). How can public health concerns balance with personal and economic rights? Who and how do governments set the development goals? What powers do governments at different levels have? Different from natural disasters in history, coping with the pandemic disruptions is far beyond an economic and health issue. At the same time, due to a wide variety of factors, many businesses may not take the opportunities provided by government aid programs. Especially, for the socioeconomically vulnerable business communities, the difficulties in application processes, governmental distrust, and the confusion of

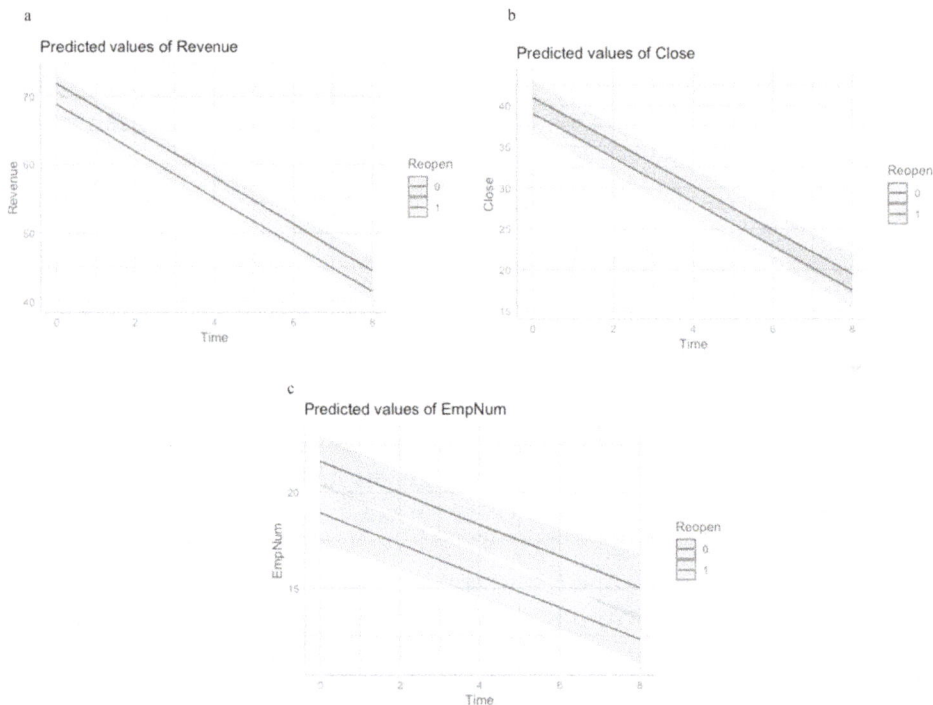

FIG. 11—Predicted values of (a) revenue reduction, (b) business temporary close, and (c) reduction of employment with reopen over time.

rules just made them less likely to take part in assistance programs than their nonminority, small-business counterparts (Bartik and others 2020).

CONCLUSION

This study contributes to our understanding of the impacts of COVID-19 on small businesses. Built upon the literature of small-business resilience and regional resilience, this study investigates multiple impacts of COVID-19 on small businesses in the most-populous 50 MSAs in the United States. The studied MSAs have shown significant differences in pandemic impacts. Our study also highlights that the pandemic disruption on small businesses was highly volatile and changed significantly each week from April 26 to June 27, 2020. Due to limited data availability, we are not able to know how representative our findings are of the entire nation. However, as shown by the recently released 2015–2019 ACS, the 50 most-populous MSAs were home to 54.9 percent of the U.S. population. The 2018 County Business Patterns data shows that for every 100 small businesses in the United States, about 58 were located at one of the 50 most-populous MSAs. In addition, for every 100 small businesses located at a MSA, about 67 were located at one of the 50 largest MSAs. Therefore, this study is still meaningful and relevant to national patterns.

a

Predicted values of Revenue

b

Predicted values of EmpNum

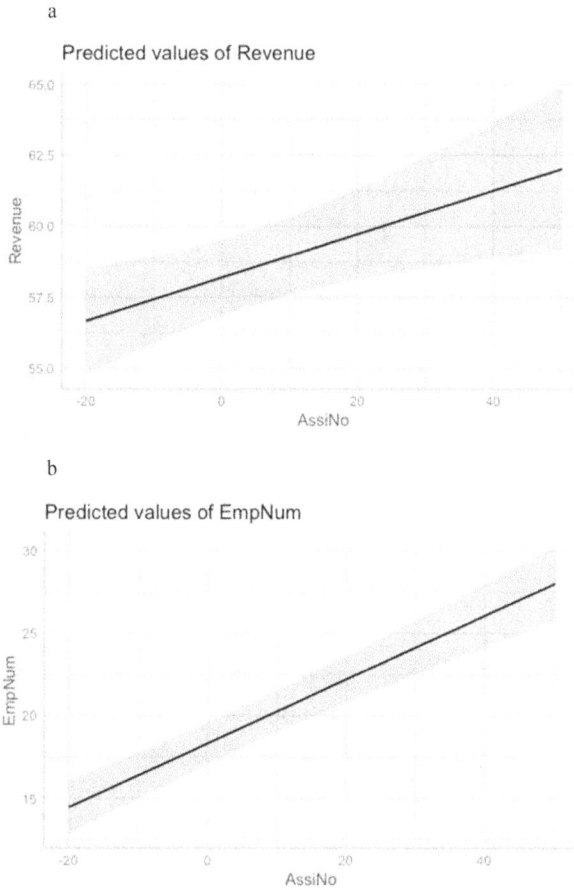

Fig. 12—Predicted values of (a) revenue reduction and (b) employment reduction with the increase of non-aided proportion in a MSA.

Different from the regional resilience literature, industrial diversification and sectoral mix are not significant factors associated with the impacts. Meanwhile, unsurprisingly, with social distancing and business shutdown, the death-case incidence and virality of infection are associated with reduced business operations, as well as the pessimistic perception of the recovery. Importantly, a region's economic, social, and demographic vulnerability is significantly associated with small businesses' reduction in revenue and operation, as well as a negative perception of recovery and overall impacts of COVID-19. Finally, public policies are significantly related to the impacts on business operations. For example, reopening the economy directly increases revenues or sales, the size of employment, and working hours for small businesses, as well as reduced-business temporary closures. Furthermore, more federal business assistance in a region is positively associated with less business shutdown, an increase in employment and working hours, but a negative expectation of recovery.

Although a long-term perspective is necessary to better understand regional resilience in the context of the COVID-19 shock, a closer examination of the current short-term impacts across regions helps to formulate more localized and targeted policy proposals. Small-business owners are particularly impacted when disaster strikes—they are impacted as business owners, and also as private citizens (Runyan 2006; Haynes and others 2019). This is underscored by the inequalities disasters lay bare, for example in resources, information, and the ability to act; as such, the socioeconomic characteristics impact their ability to prepare for, anticipate, cope with, and recover from hazard events (Peacock and others 2014). Extant literature has also long suggested that the spatial concentration of poverty, social exclusion, and racial segregation are the root cause of spatial patterns of health inequality. In considering that businesses in these underserved communities have already faced more barriers for start-up and survival (Bates and others 2018), the pandemic leaves these communities more vulnerable to disproportionate exposure and more difficult to recover postdisaster. Therefore, policy and practices should target the socioeconomically vulnerable communities from an interconnected perspective and simultaneously consider the health, economic, and social impacts.

In this study, we only present an analysis of COVID-19 impacts based on a limited number of metropolitan areas, measuring a short term of nine weeks. With the economy reopening and the volatile changes of COVID-19, more observations and long-term investigation are needed. At the regional level, further studies on the typical process of regional resilience, which are long-term in character and focus on adaptability and qualitative changes, need to be investigated across different regions. In particular, how each region recovers, reorients, and reorganizes, and how this process will impact small businesses' recovery, adaptation, and transformation needs long-term, place-specific, contextualized analyses. We particularly need microdata to examine the experiences of individual firms, in terms of how they cope with the crisis, survive, and innovate. Furthermore, more research should be conducted to treat small-businesses' coping process as a dynamic interaction and collaboration between individuals, households, businesses, and communities to adapt to external shocks and ultimately survive and thrive.

FUNDING

This work was supported by the NSF [2029516].

NOTES

[1] It should be noted that as the COVID-19 cases continued to surge, some states have changed their reopen policies back and forth since the first lockdown in 2020.

[2] Currently the survey does not provide firm level data, but only aggregated data at the national, state, and the largest 50 MSA level. The survey design, questionnaire, data, and relevant

documents can be obtained here https://www.census.gov/data/experimental-data-products/small-business-pulse-survey.html

[3] The HHI for MSA i is formally defined in the following Equation: $HHI_i = \sum_{s=1}^{k} (\frac{E_{si}}{E_i})^2$ where E_{si} is the employment of industry s in MSA i, E_i is the total employment in MSA i, and k is the number of North American Industry Classification System (NAICS) industrial sectors. We rely on the two-digit NAICS code system (k = 20) due to limited data availability.

[4] The CDC provides the SVI data at the census tract and county level. Since our unit of analysis is MSA, we aggregate the county-level SVI data to the MSA level using the population-weighted average technique.

ORCID

Wei Kang ⓘ http://orcid.org/0000-0002-1073-7781

REFERENCES

Amram, O., S. Amiri, R. B. Lutz, B. J. Rajan, and P. Monsivais. 2020. Development of a Vulnerability Index for Diagnosis with the Novel Coronavirus, COVID-19, in Washington State, USA. *Health & Place* 64 (2020):102377. doi:10.1016/j.healthplace.2020.102377.

Bartik, A. W., M. Bertrand, Z. Cullen, E. L. Glaeser, M. Luca, and C. Stanton. 2020. The Impact of COVID-19 on Small Business Outcomes and Expectations. *Proceedings of the National Academy of Sciences* 117 (30):17656–17666. doi:10.1073/pnas.2006991117.

Bates, T., W. D. Bradford, and R. Seamans. 2018. Minority Entrepreneurship in Twenty-First Century America. *Small Business Economics* 50 (3):415–427. doi:10.1007/s11187-017-9883-5.

Beaunoyer, E., S. Dupéré, and M. J. Guitton. 2020. COVID-19 and Digital Inequalities: Reciprocal Impacts and Mitigation Strategies. *Computers in Human Behavior* 111:106424. doi:10.1016/j.chb.2020.106424.

Crescenzi, R., D. Luca, and S. Milio. 2016. The Geography of the Economic Crisis in Europe: National Macroeconomic Conditions, Regional Structural Factors and Short-Term Economic Performance. *Cambridge Journal of Regions, Economy and Society* 9 (1):13–32. doi:10.1093/cjres/rsv031.

Di Caro, P., and U. Fratesi. 2018. Regional Determinants of Economic Resilience. *The Annals of Regional Science* 60 (2):235–240. doi:10.1007/s00168-017-0858-x.

Donthu, N., and A. Gustafsson. 2020. Effects of COVID-19 on Business and Research. *Journal of Business Research* 117:284. doi:10.1016/j.jbusres.2020.06.008.

Evenhuis, E. 2017. New Directions in Researching Regional Economic Resilience and Adaptation. *Geography Compass* 11 (11):e12333. doi:10.1111/gec3.12333.

Fairlie, R. 2020. *The Impact of COVID-19 on Small Business Owners: Evidence of Early-Stage Losses from the April 2020 Current Population Survey*, w27309. Cambridge, MA: National Bureau of Economic Research. doi:10.3386/w27309.

Flanagan, B. E., E. J. Hallisey, E. Adams, and A. Lavery. 2018. Measuring Community Vulnerability to Natural and Anthropogenic Hazards: The Centers for Disease Control and Prevention's Social Vulnerability Index. *Journal of Environmental Health* 80 (10):34.

Fox, J. 2015. *Applied Regression Analysis and Generalized Linear Models*. Thousand Oaks, CA: Sage Publications.

Giannakis, E., and A. Bruggeman. 2017. Determinants of Regional Resilience to Economic Crisis: A European Perspective. *European Planning Studies* 25 (8):1394–1415. doi:10.1080/09654313.2017.1319464.

Golden, T., and M. L. Wendel. 2020. Public Health's Next Step in Advancing Equity: Re-Evaluating Epistemological Assumptions to Move Social Determinants from Theory to Practice. *Frontiers in Public Health* 8:131. doi:10.3389/fpubh.2020.00131.

Gong, H., R. Hassink, J. Tan, and D. Huang. 2020. Regional Resilience in Times of a Pandemic Crisis: The Case of COVID-19 in China. *Tijdschrift Voor Economische En Sociale Geografie* 111 (3):497–512. doi:10.1111/tesg.12447.

Gostin, L. O., and L. F. Wiley. 2020. Governmental Public Health Powers during the COVID-19 Pandemic: Stay-at-Home Orders, Business Closures, and Travel Restrictions. *Jama* 323 (21):2137–2138. doi:10.1001/jama.2020.5460.

Haynes, G. W., S. M. Danes, H. L. Schrank, and Y. Lee. 2019. Survival and Success of Family-Owned Small Businesses after Hurricane Katrina: Impact of Disaster Assistance and Adaptive Capacity. *Journal of Contingencies and Crisis Management* 27 (2):130–144. doi:10.1111/1468-5973.12245.

Kakderi, C., and A. Tasopoulou. 2017. Regional Economic Resilience: The Role of National and Regional Policies. *European Planning Studies* 25 (8):1435–1453. doi:10.1080/09654313.2017.1322041.

Karaye, I. M., and J. A. Horney. 2020. The Impact of Social Vulnerability on COVID-19 in the US: An Analysis of Spatially Varying Relationships. *American Journal of Preventive Medicine* 59 (3):317–325. doi:10.1016/j.amepre.2020.06.006.

Kitsos, A., and P. Bishop. 2018. Economic Resilience in Great Britain: The Crisis Impact and Its Determining Factors for Local Authority Districts. *The Annals of Regional Science* 60 (2):329–347. doi:10.1007/s00168-016-0797-y.

Marshall, M. I., and H. L. Schrank. 2014. Small Business Disaster Recovery: A Research Framework. *Natural Hazards* 72 (2):597–616. doi:10.1007/s11069-013-1025-z.

Martin, R., and P. Sunley. 2015. On the Notion of Regional Economic Resilience: Conceptualization and Explanation. *Journal of Economic Geography* 15 (1):1–42. doi:10.1093/jeg/lbu015.

Martin, R., P. Sunley, B. Gardiner, and P. Tyler. 2016. How Regions React to Recessions: Resilience and the Role of Economic Structure. *Regional Studies* 50 (4):561–585. doi:10.1080/00343404.2015.1136410.

Nowell, B., C. P. Bodkin, and D. Bayoumi. 2017. Redundancy as A Strategy in Disaster Response Systems: A Pathway to Resilience or A Recipe for Disaster? *Journal of Contingencies and Crisis Management* 25 (3):123–135. doi:10.1111/1468-5973.12178.

NYT (*New York Times*). 2020a. Loan Money Runs Out while Small-Business Owners Wait in Line. April 16. https://www.nytimes.com/2020/04/16/business/coronavirus-sba-loans-out-of-money.html.

_____. 2020b. Small-Business Loan Program, Chaotic from Start, Gets 2nd Round. April 26. https://www.nytimes.com/2020/04/26/business/ppp-small-business-loans.html.

_____. 2020c. Bankers Rebuke S.B.A. As Loan System Crashes in Flood of Applications. April 27. https://www.nytimes.com/2020/04/27/business/sba-loan-system-crash.html.

Palekiene, O., Z. Simanaviciene, and J. Bruneckiene. 2015. The Application of Resilience Concept in the Regional Development Context. *Procedia - Social and Behavioral Sciences*, 20th International Scientific Conference "Economics and Management 2015 (ICEM-2015)," 213 (December):179–184. doi:10.1016/j.sbspro.2015.11.423.

Peacock, W. G., S. Van Zandt, Y. Zhang, and W. E. Highfield. 2014. Inequities in Long-Term Housing Recovery after Disasters. *Journal of the American Planning Association* 80 (4):356–371. doi:10.1080/01944363.2014.980440.

Ruggles, S., S. Flood, R. Goeken, J. Grover, E. Meyer, J. Pacas, and M. Sobek. 2020. *IPUMS USA: Version 10.0[dataset]*. Minneapolis, MN: IPUMS. doi:10.18128/D010.V10.0

Runyan, R. C. 2006. Small Business in the Face of Crisis: Identifying Barriers to Recovery from a Natural Disaster 1. *Journal of Contingencies and Crisis Management* 14 (1):12–26. doi:10.1111/j.1468-5973.2006.00477.x.

Sedita, S. R., I. De Noni, and L. Pilotti. 2017. Out of the Crisis: An Empirical Investigation of Place-Specific Determinants of Economic Resilience. *European Planning Studies* 25 (2):155–180. doi:10.1080/09654313.2016.1261804.

Seetharaman, P. 2020. Business Models Shifts: Impact of COVID-19. *International Journal of Information Management* 54:102173. doi:10.1016/j.ijinfomgt.2020.102173.

Tibay, V., J. Miller, A. (Y.) Chang-Richards, T. Egbelakin, E. Seville, and S. Wilkinson. 2018. Business Resilience: A Study of Auckland Hospitality Sector. *Procedia Engineering* 212:1217–1224. doi:10.1016/j.proeng.2018.01.157.

Torres, A. P., and M. I. Marshall. 2015. Does Social Capital Explain Small Business Resilience? A Panel Data Analysis Post-Katrina. 205080. 2015 AAEA & WAEA Joint Annual Meeting, San

Francisco, CA: Agricultural and Applied Economics Association, July 26–28. https://ideas. repec.org/p/ags/aaea15/205080.html.

Tsai, J., and M. Wilson. 2020. COVID-19: A Potential Public Health Problem for Homeless Populations. *The Lancet Public Health* 5 (4):e186–87. doi:10.1016/S2468-2667(20)30053-0.

Van Zandt, S., W. G. Peacock, D. W. Henry, H. Grover, W. E. Highfield, and S. D. Brody. 2012. Mapping Social Vulnerability to Enhance Housing and Neighborhood Resilience. *Housing Policy Debate* 22 (1):29–55. doi:10.1080/10511482.2011.624528.

Ventres, W. 2017. From Social Determinants to Social Interdependency: Theory, Reflection, and Engagement. *Social Medicine* 11 (2):84–89.

Wenham, C., J. Smith, and R. Morgan. 2020. COVID-19: The Gendered Impacts of the Outbreak. *The Lancet* 395 (10227):846–848. doi:10.1016/S0140-6736(20)30526-2.

Wisner, B., P. Blaikie, T. Cannon, and I. Davis. 2003. *At Risk: Natural Hazards, People's Vulnerability and Disasters*. 2nd ed. London and New York: Routledge.

Xiao, Y., and W. G. Peacock. 2014. Do Hazard Mitigation and Preparedness Reduce Physical Damage to Businesses in Disasters? Critical Role of Business Disaster Planning. *Natural Hazards Review* 15 (3):04014007. doi:10.1061/(ASCE)NH.1527-6996.0000137.

MASKS AND MATERIALITY IN THE ERA OF COVID-19

SARA BETH KEOUGH●

ABSTRACT. During the COVID-19 pandemic, cloth and disposable masks were recommended to reduce virus transmission. As mask wearing became more common, whether by personal choice or regulation, and the variety of masks available increased, these masks acquired the status of key material-culture objects. This essay reflects on the symbolic nature of obtaining, wearing, and discarding masks during the pandemic within the context of Michigan in the United States. Using material-culture theories proposed by Anna Tsing, Arjun Appadurai, Christine Harold, and Andrei Guruianu and Natalia Andrievskikh, I consider the cultural value, power, and communicability of masks. Woven into this reflection are my personal experiences with masks as they relate to the three themes. I conclude by reflecting on the ephemeral nature of masks and the ever-changing meanings derived through an examination of material culture.

A day or two prior to the start of each semester, I find myself putting too much time into planning what I am going to wear on the first day. I stare into my closet wondering what I want my clothing choice to communicate to students. I think, the suit? No, it will make me seem too stiff and I want my students to feel comfortable enough to participate and share ideas. Nice jeans and shirt? No, that could appear too casual and this class is a challenging one that the students need to take seriously. I've even joked with my fellow faculty about showing up on the first day in my doctoral robes, Professor McGonagall-style, a reminder to students that I have earned the degrees that come with the title "Dr." or "Professor," and an attempt to discourage their use of "Mrs." and other prefixes that refer to a presumed marital status. This year, however, it wasn't the clothing, but my mask that I was contemplating.

My university, like many others, spent the summer establishing a plan for students, faculty, and administrators to safely return to campus in the fall. Among other things like reduced classroom capacities, copious amounts of disinfectant wipes and hand sanitizer in all rooms and common areas, and virtual office hours, the university mandated that masks must be worn at all time inside any campus building and outside as well if anyone is within six feet of another person. The Office of Alumni Affairs provided one cloth mask with the university mascot to each faculty and staff member on campus, as well as to students who were paid to help enforce these policies by encouraging mask wearing, providing masks to those not wearing one, and setting a good example by wearing one themselves. Initially, however, this was not the mask I chose to wear on the first day.

My mask-providing university is based in Michigan, a state whose female Democratic governor was lauded by some and criticized by others, President Trump in particular, for the strict measures on commerce, social gatherings, and mask wearing that she implemented between March and October 2020 using executive orders.[1] As a material-culture scholar and human geographer, I am forever curious about the value held by material objects obtained, displayed, or discarded. The arrival of COVID-19 in early 2020, and in my state in March of that year, presented new material cultures that communicated local and national values, preferences, and identities. In Michigan, toilet paper, disinfecting wipes, hand sanitizer, and raman noodle packages became objects valuable far beyond their market price as they disappeared from the literal and virtual shelves of stores from Kroger to Amazon, were hoarded by some, and suddenly rationed by those who had only a few weeks-worth stored away. Yet, to me, mask distribution, buying, and wearing stuck out because these actions were done in mostly public ways. Whether or not one wears a mask beyond private spaces, and what kind of mask they wear or what is depicted on the mask, communicates perspectives on the pandemic, elements of American cultural values, personal identity, and even politics, as responses to the pandemic in the United States are as much a political issue as a public health one.

This essay is a reflection on masks as material culture in the era of COVID-19. It is a collection of my observations and thoughts as I lived through the pandemic in Michigan. My reflections are informed by my own expertise as a scholar of material culture, in addition to material-culture theorists like Anna Tsing,, Arjun Appadurai, Christine Harold, and Andrei Guruianu and Natalia Andrievskikh. These reflections are divided into three parts, though they are not mutually exclusive: obtaining masks, where I reflect on the perceived value of the mask itself in its various forms; wearing masks, where I reflect on the political and personal identities and powers communicated through the act of wearing (or not) a mask during the pandemic; and discarding masks, where I consider the irony in the ephemeral nature of masks. This essay is not meant to be a comprehensive review of masks as material culture, as their meaning and value is changing even as I write this and will likely evolve further before this essay is published. Instead, my goal is to begin an interpretation of the mask within the context of a global pandemic and local—Michigan—cultural responses that can be expanded on as this phenomenon continues.

OBTAINING MASKS

In May 2020, the Organization for Economic Cooperation and Development (OECD) released the results of a study on the global value chain of masks designed to protect the wearer against COVID-19 (OECD 2020).[2] This study, which focused only on the disposable N95 "surgical" or "medical" masks and respirators, traced the commodities involved in production, dependent, of course, on the global market value of these commodities, as well as costs related to production, sterilization, and the careful packaging required of a sterilized product. While global gas prices plummeted as stay-at-home orders were issued in countries across the world, the demand for these polypropylene (a polymer derived from oil) face masks skyrocketed. National Public Radio reported that in March 2020 in the United States, a pack of 30 N95 masks was selling on Amazon for US$199, when the average price prior to the pandemic was around US$15. Delivery time was more than a month (Rosalsky 2020). Interrupted global supply chains, labor shortages in many industries, price gouging by companies selling masks, and hoarding by those who could afford to buy them accentuated the problem. Then, of course, the Centers for Disease Control and Prevention (CDC) message that the mask, as a form of PPE, could save one's life fed the demand. In short, masks became highly valuable objects exchanged in a capitalist system.

When school closures and shutdowns were first announced in mid-March 2020, and stay-at-home orders issued, finding masks was a challenge. Most of the masks available were disposable. Some people bought large quantities and hoarded them. Three-ply and surgical (N95) masks were the most common, but we were encouraged to save them for essential workers. It took a while for the U.S. industrial sector to mobilize in mask production. There was also debate over which kind of mask was most effective. The N95 masks are still considered one of the best lines of defense if one must engage in public spaces, and in Michigan we were encouraged to "save" those masks (as in, not buy them and/or hoard them) for those who could not quarantine due to the essential nature of their jobs. Hand- or machine-sewn cloth masks, which unlike N95 masks were reusable, had to fill the gap, and in my community, several groups mobilized to make and distribute cloth masks to as many people as possible. Demand for masks was high, supply was low. The lost art of sewing reemerged on the (quarantined) domestic landscape. Not only were masks highly valued as a form of personal protective equipment (PPE), but the commodities and tools required to make them, and individuals who garnered this knowledge, became, overnight, highly valued elements of society.

I obtained my first mask in mid-March 2020 from my neighbor, as a gift. I did not pay for the mask itself or the supplies used to make it. My neighbor, an avid sewer, a retired postal worker, and a person genuinely concerned about her fellow human beings, began sewing masks using a pattern she found online (with CDC recommendations for its construction). I had been unsuccessful at that point in securing a mask for myself and my elementary school–aged son and was growing concerned. She didn't know this, however. She texted me one afternoon in mid-March, just before Michigan's stay-at-home orders were issued, and said she had just left three masks in a plastic bag on my front porch. One for me, one for my son, and one for my former spouse (a person she had never met). She made my mask out of a cotton pattern depicting Van Gogh's *Starry Night* because she knows I "do things with the arts," and my son's mask out of Michigan State University cloth, his favorite in-state sports team. These masks were both highly valuable commodities and gifts to us.

Anthropologist Anna Tsing writes about the differences and connections between gifts and commodities (Tsing 2013, 2015). According to Tsing (2013), in a capitalist economy, "commodities *define* the system" [emphasis in original]. Taking the gift out of the commodity is very difficult and the two are not mutually exclusive categories. There are some distinctions, however. The degree of alienation between recipient and producer is greater for commodities than for gifts, or for commodities that have become gifts. Furthermore, "value in a commodity system is in things for use and exchange," whereas in a gift system, it is "social obligations, connections, and gaps" between players that add value (Tsing 2013, 22).

The mask from my neighbor was inarguably a gift. Although I did not know the cotton farmer and the fabric was probably woven by machine, I did know the person who transformed those resources into my mask. The degree of alienation was minimal. The gift of the mask connected us and bridged a social-interaction gap that lockdown orders had created. At the same time, however, my gifted mask was also a commodity. I had something that other people were desperately trying to find, sometimes spending large amounts of money to obtain. Its exchange value, had I chosen to sell it, was considerable. Its use value even more so. The mask served as both a protective agent and the means to obey state laws, as it enabled me to make occasional grocery store runs or get something I needed from my campus office. When the same neighbor that gifted me the mask got sick, she asked me if I could pick up a prescription for her and leave it on her porch, a task I was willing and able to do because I now had a protective mask.

The difference between my first mask and the ones people were combing Internet sites to purchase is fairly clear. People trying to buy masks—or any other commodities in high demand, like toilet paper, for example—were not concerned with who made it. There was no relationship with the producer or any part of the supply chain prior to or after the product was obtained by the person who sought it. Hoarding strictly involved commodities, as those with an abundance likely kept them for themselves, friends, or family, or resold the surplus at higher prices. Although there are clear relationships between gifts and commodities, my first mask was clearly a gift, and not just because I didn't pay for it. Our masks were made by my neighbor specifically for me and my son. We didn't ask for one. She chose the cloth to reflect our interests. She took care to place them in Ziploc bags and not come into contact with us upon delivery. They were personal and contextual. They further solidified our relationship as friends because the act of giving them expressed a genuine concern on the part of my neighbor for our well-being. It was a warm and selfless dyadic gesture that came during a consumer period dominated by hoarding.

It is not, however, the only mask I have that I didn't pay for. My university gave each faculty member a single mask with a small version of the mascot on the side, as they required mask wearing to enter campus. Even though I didn't have to pay for this mask, I don't consider it a gift. Each faculty member (and staff and administrators, and most students, and several alumni) received the same mask, so there was no personalization. Masks were delivered to our mailboxes and mine greeted me when I returned to my campus office after not having gone to campus for more than three months. I did not know when it arrived, how long it had been there, or who placed it in that location. It was contained within a commercially sealed package. The degree of alienation between myself and my university-provided mask is great. I have no personal attachment to this mask, even though it bears my university's mascot.

Anna Tsing (2013) warns us about creating dichotomies between gifts and commodities because systems of exchange are "mixed and messy." Designation as a gift or a commodity is often based on the perception of the person doing the categorizing. The same object can be a gift to one person and a commodity to another, switch back and forth between these two designations, or exist as both at the same time. Most gifts, including my Van Gogh mask, are made from raw materials and depend on global supply and value chains. As described earlier, my Van Gogh mask was a commodity first, and a gift second. Although I do not know where the cotton to make this mask was grown and harvested, where the fabric was woven and imprinted, and where the elastic was produced, I do know who sewed the mask, how and why she chose the fabric, and by what means and intentions it was delivered. Furthermore, the Van Gogh mask facilitated a stronger relationship between myself and my neighbor in part because of this knowledge about its production and intended purpose.

On the other hand, my university-provided mask remained a commodity even after it was in my possession. A few weeks later, however, after discovering how comfortable it was and that I could wear it without causing my glasses to fog up, I asked the Office of Alumni Affairs (which had paid for, ordered, and distributed the masks on campus) if there were any left, as I had a friend who struggled to wear glasses and a mask at the same time in her job. The director gave me five masks in the same commercially sealed packages, which I then gave to friends. Were these gifts? I gave them to people with whom I already had a strong personal relationship. I did not make the masks myself, nor did I pay for them. To my friends, the masks were objects that facilitated their work. I am not sure if the actual mask itself held any emotional value, though my friends probably appreciated the gesture.

Arjun Appadurai (1986, 2012), another scholar of material culture, rejects the gift-commodity distinction. Appadurai sees anything that is exchanged as a commodity because it has economic value. My Van Gogh mask had/has economic value. I could have sold it for probably US$30 back in May. My university-provided mask does not lack economic value just because I did not pay for it. I could have sold it as well. It is still the most effective mask I have for teaching because it is the only one that doesn't fog up my glasses, and that factor likely increases its value as a commodity. In exchange for the mask from my university (and, of course, my salary), I am expected to deliver content to students in person because I have been provided with a relatively safe means for doing so.

As of this writing, the costs to consumers of masks have dropped substantially. A quick search on Amazon (October 2020) shows that N95 masks are running around US$1.50 each, or a pack of 20 for about US$30. The general 3-ply masks are selling for around US$0.50 each, or a pack of 50 for US$25. Both kinds qualify for Amazon Prime two-day shipping from several suppliers. My Van Gogh mask, which I would have considered a gift at any point, was also valuable in the timing of its deliver. In other words, the processes for obtaining masks and their commodity value have changed drastically over the last year. In February 2021, I noticed individual masks available free at a local Home Depot. In one year, these masks had lost their entire value. With more availability comes greater consumer choice. Now, instead of using just any mask we can get, we can choose to wear masks that convey messages and identities.

WEARING MASKS

For my first-day-of-school mask, I went with practical. I chose a white cotton mask I had been given by the United States Census Bureau to use while working as a census enumerator over the summer. It was comfortable, allowed me to breathe sufficiently, and it was a neutral color. It did not, however, prevent my glasses from fogging up, which is what prompted me to try the university-provided mask. In anticipation of the semester, and teaching partly in person, I had also purchased two masks from Amazon on which a map of the world was printed. I am a geography professor and a map lover. These masks were an expression of one part of my personal identity. When they arrived in the mail, however, I discovered the masks were made of multiple layers of silk and were thick. Silk is a comfortable fabric for sure, but the multiple layers made it almost impossible to take deep breaths, something I have to do often while standing and talking at the front of a big classroom trying to project to students spread out as much as possible.

This process of choosing and then wearing a mask, and then observing the mask choices of my students, colleagues, and the general public, made me think more seriously about mask wearing as a form of identity expression. For me, masks went from being objects to being subjects when personal choice became part of process. One of the ways a material object becomes an expression of identity is when the person using it for this purpose has a variety of types, brands, or patterns from which to choose.

By the time the Fall 2020 semester started, the supply of masks in the United States had caught up to demand. Disposable masks were readily available everywhere from Amazon to the nearby Kroger. Facebook advertisements for masks in varying designs and fabrics (mostly reusable ones) were regularly infiltrating my news feed. In other words, masks could be an expression of one's identity because there were now thousands of prints and tens of styles to choose from. A person's mask went from being anything they could find that might work as a protective layer to an act of personal choice and expression of identity.

Perhaps one of the drivers behind masks as expressions of personal identity was that when they were initially recommended and in high demand, the masks available were one size fits all, or one type fits all. When masks were largely unavailable or inaccessible, consumers had little choice if they actually found one. This supply/demand problem led to objects turned into masks that were not initially intended for that purpose, things like a shirt or neck warmer pulled over one's nose or a handkerchief held up to one's face.

With time, patterns used in cloth masks made for others also became a way for mask makers to communicate sentiments toward those receiving the masks. Mask-sewing groups chose various fabrics and prints as they made masks initially for healthcare workers experiencing a shortage of PPE, and then for other groups in the public. One CNN report showed images of hearts and flower patterns used by a group in Atlanta, an expression of appreciation for the sacrifice healthcare and other essential workers were making (CNN 2020). The Facebook group "Sewing Mask Pattern," which according to CNN had just over 3000 members in March 2020, now shows 12.2 thousand members January 2021. The group "Mask Makers Community" has 13.9 thousand members.

The act of producing this material culture falls, of course, in the context of a global pandemic and a shortage of protective equipment. As the country waited for an already reduced industrial sector to mobilize and meet demand, these grassroots sewing groups stepped up in World War II collective fashion. The key difference, however, is that while WWII sewing groups involved a physical mobilization of women in groups together in the same space, the COVID-19 mask-sewing groups were, to use Benedict Anderson's (1983) term, largely imagined communities.[3] They emerged as concerns about the availability of PPE became louder and alarming, and were formed largely via social media from neighborhoods, friends, church groups, and other entities that were established pre-COVID, usually for entirely different purposes. People sewing masks during COVID-19 were working alone, at home, by themselves, often using material they already had, as fabric stores were closed in the early days of the pandemic. Instead of sharing techniques and patterns in person, people sewing masks used YouTube videos for instruction on how to design the most effective ones. In a sense, these YouTube videos were gifts themselves, free classes from those who had the knowledge and skill to those who needed it for a greater purpose.[4]

As mask wearing became more ubiquitous in public spaces and buildings, and the necessity of mask wearing more urgently communicated to the general public, so did the presence of the face mask in advertising. The company Phantom Fireworks and its highway billboards, which tend to proliferate the advertising space along Midwest interstate highways during the weeks before July 4th, this year depicted the phantom with a mask on. The Michigan State University Spartan statue wore a mask as early as April 2020 ("Michigan State's Sparty Statue ..." 2020). By late summer 2020, television ads included actors doning face masks. In a pre-Halloween 2020 Facebook post, a friend from graduate school posted a picture of the pumpkin on the front step of his house wearing a facemask. By January 2021, television dramas like *NCIS: Los Angeles*, which had begun filming new episodes, had their main characters donning masks when interacting with the public. In other words, wearing a face mask is slowly (too slowly, in my opinion) becoming commonplace.

Although mask wearing was slow to catch on in the United States, one reason that the practice had relatively slow adoption rates in Michigan, in particular, was because wearing a mask became a political expression. When the pandemic hit Michigan, Governor Gretchen Witmer issued stay-at-home orders and man-dated mask wearing (a mandate that is still in place as of this writing) through the power of executive order, largely because general recommendations to wear masks and socially distance were initially ignored by many. Witmer's use of executive order was controversial and sparked large protests; the largest one held in Lansing on 1 May 2020 attracted national and international media coverage (Slotkin 2020). At least three people were killed in confrontations at businesses attempting to enforce the mask-wearing mandate, and some law enforcement agencies in the state refused to enforce the order. Yet 92 percent of Michiganders surveyed as part of the Democratic Fund + ULCA Nationscape Project (15 June–11 July 2020) said they had worn a mask when they went out in public. Nationally, the response was 89 percent (Kovanis 2020). In other words, there was some discrepancy between what people say they do—and how they respond to a survey question when they know the answer they should give—and what they actually do. Images of protesters at the state capitol throughout the spring, summer, and fall showed these groups gathering without masks and social distancing, in defiance of the executive order given by the governor (Oosting and Beggin 2020; Slotkin 2020). Refusing to wear a mask in public became synonymous with defiance of the governor's order and declara-tion of independence and freedom, while wearing a mask and practicing social distancing came to be viewed as a sign of support, regardless of whether or not the individual was intentionally taking a political stance.[5]

Politics aside, the perception of the mask's power to protect is one worth mentioning. Using photographs posted to Facebook by my "friends" as an example, I see a group of people standing much closer than six feet, but all wearing masks. To me, this is an acknowledgment of the power of masks to protect, especially in situations where people are closer than the recommended distance. This image, and one where people are standing six feet apart and are also masked says "we are doing our part to protect others" and "we believe these measures will help us." Although CDC guidelines indicate the six-foot distance and mask wearing both, there is the perception that masks have the power to protect when social distancing is not possible. The perceived protective power of the mask reduces space, where not wearing one among concerned individuals often results in increased physical distance between them. In this way, masks can empower the wearer to take some greater risks than that person might have been willing to take without the mask.

There is a point, however, where the mask becomes a dangerous object. Concern about how often they are washed between uses, and whether or not they have been touched by hands that have come in contact with contaminated surfaces grows. The ubiquitous problem of mask slippage means I am constantly adjusting my mask on my face, and then trying to remember when I last used sanitizer or washed my hands. Here, the materiality of the mask changes. What was once an agent of protection becomes a potential agent for contagion. Unlike the KonMari Method, the currently popular method for tidying up designed by Marie Kondo that encourages the individual to hold an object in their hands, experience its tangibleness, and decide whether the object produces joy, touching a mask makes it a more dangerous object, regardless of the connection to it felt by the wearer. Furthermore, the inability of the face mask to be 100 percent protective, thus negating their usefulness, has been exploited by antimaskers.

Masks also have the power to hide. Popular culture icons like Batman, Zorro, and the Lone Ranger, to name just a few, wore masks that appeared to make them unrecognizable to the people they came in contact with. Batman's mask was full-face, Zorro's and the Lone Ranger's just covered their eyes, but left their mouth exposed. Batman's mask was also protective, Zorro's and the Lone Ranger's just hid their identity. None of these masks were used as a protective health device, of course. And they were all black. Although the eyes can communicate emotion, the viewer is left with few other clues to a mask-wearer's reaction, at least fewer clues than we are used to having. One *New York Times* article that was widely circulated in my academic circles described how African-American men were concerned about the increased tendency for racial profiling brought about by mask-wearing mandates (Taylor 2020). More so than other groups, African-Americans, and African-American men in particular, felt forced to choose between their health and their safety, citing concerns voiced by non-Blacks that African-Americans in masks were "up to no good" (Taylor 2020). Whether a mask hides the eyes or the mouth can change the perception of the viewer, as well as any racial profiling the viewer may be guilty of due to the color of the wearer's skin, their size, their gender. Masks also prevent the wearer from using most facial expressions to convey sentiment, something I feel limits me in the classroom while I teach with my mask on.

Thus, masks are material culture that exist in this marginal space between law and personal choice, between individual protection and community concerns, between hidden identities and outward displays of personal identities, between empowerment and disempowerment, between safety and danger. Yet, we cannot end the mask discussion without briefly considering what happens to these material items after we wear them.

DISCARDING MASKS

Communication Professor Christine Harold (2020, 4) calls waste "the pejorative term for excess," a condition generated by perceived and planned obsolescence ubiquitous in capitalist economies. Waste implies a growing consumer culture and has devastating environmental impacts (Harold 2020). In March of 2020, there was hardy an excess of masks, so the idea of disposable masks was striking. While I certainly understood the public health reasons for discarding a mask after one use, the idea of doing so was difficult, knowing masks were limited in number. Still, I saw these disposable masks strewn across the Target parking lot in May 2020. Recycling anything came to a halt in Michigan, as curbside recycling pickup was suspended and bottle returns closed. Waste in general accumulated, and discarded masks and gloves, which could not be recycled, added to the volume. In "tragedy of the commons" style, human health was prioritized over environmental health (Adyel 2020).

Masks in the COVID-19 era maintain an ephemeral nature because health officials recommended against reuse. As production of disposable masks caught up to demand, and disposable masks became much easier to obtain, their life span dwindled further. An experience outside my local Target store is worth mentioning. In Michigan, as of this writing, mask wearing is required indoors in all buildings outside of one's private residence. When someone puts on a mask because it is required to, say, enter a store, the value of the mask is held in its protective capabilities: both protection from COVID-19 and protection from being asked to leave the store. (I'm speaking mostly about disposable masks here, as that is what I observe a majority of shoppers and employees wearing.) That value is immediately lost in the parking lot, however, because the requirement to wear it is no longer in place. The risk of being rejected from the store as a customer is eliminated, and the risk of contacting COVID-19 in the open-air parking lot is significantly reduced. The act of discarding the mask on the ground, instead of in the waste bin at the store exit, further drives home for me not only that value has been lost, but how quickly that protective object became valueless—it wasn't even worth depositing in the trash bin. The presence of disposable masks laying on the ground in a puddle is visible evidence that their purpose has expired. It could also be seen as a communicative act by those who object to mandated mask wearing.

In *The Afterlife of Discarded Objects* (2019), authors Guruianu and Andrievskikh explore how disposed materials become "something more" after their initial purpose has expired. An old ticket stub becomes a memory, and family object becomes an heirloom as it gets passed through generations, discarded plastic bags get made into kites—these examples demonstrate added value and life to objects over time. Will discarded masks have an afterlife? At the moment, I doubt it. A discarded mask is considered contaminated. On one side it contains the wearer's germs and on the other the germs of anyone or any environment the wearer came in contact with. COVID-19 masks have no value as shared objects. Even when they were limited in supply, stealing a used one had no benefit, as the mask's perceived contamination made it valueless after its single use had been achieved.

On the other hand, cloth masks may have an afterlife. I will likely save my Van Gogh mask from my neighbor in my box of keepsakes, as it was a gift and will serve as a memory of life during the COVID-19 pandemic. It is also possible that cloth masks may be repurposed after they are no longer needed or required, probably not by anyone other than the person who wore it, but it could be used to wrap or protect a fragile object, thereby extending its protective qualities that initiated its acquisition.

CONCLUSIONS

One afternoon in early July, I sat in my car in the parking lot of the grocery store frantically combing through my purse. I eventually dumped everything out and looked through each item individually. No luck. My Van Gogh mask was gone. I always kept it in my purse, except for when I regularly washed it. I had coveted that mask. For several months, it was the only one I had because even those I had ordered online took time to arrive, the only one I received as a gift. Although I have several masks now, enough to wear a new one each day, the Van Gogh mask holds the highest personal value. It wasn't a concern over a lack of PPE that I felt. Instead, it was a sense of loss—loss of something that a friend had made by hand for me, loss of something that connected me to her. And, a loss of something that connected me to an experience.

A few weeks later, I was doing laundry and checking through pants pockets for the random tissue that would wreak havoc on my clothes if it went through the wash cycle. I reached in the pocket of a dress and pulled out ... my Van Gogh mask. I then remembered wearing it into my school office, but then upon leaving, deciding to go for a walk on the deserted campus before returning to my car. As I was outside and alone, I took off the mask and put it in my pocket. The dress remained in my laundry room for several weeks, waiting for enough other clothes of similar colors that needed washing to justify the water. It was here that I found my mask. I felt happy. It was the first time that a mask had generated joy, instead of trepidation or concern.

What will masks remind us about the COVID-19 era? Will they become artifacts in museums of medicine and material culture? What new styles, designs, and technologies will emerge in the coming weeks, months, and years? As I write this, Michigan is experiencing a resurgence of COVID-19 cases, but unlike spring of 2020, PPE is in abundance. How will that abundance change the spread of the disease? How will the value of masks change over time? Perhaps one benefit of the ephemeral nature of material culture is that meaning and value will change, and material objects will continue to provide glimpses into culture and society.

ACKNOWLEDGMENTS

I would like to thank David Kaplan for inviting me to contribute to this special issue and members of my SVSU Writing Accountability Group for their help brainstorming about the meaning of masks. Sherrin Frances, Rob Drew, Warren Fincher, and Scott Youngstedt read earlier drafts of this manuscript and provided valuable feedback. Finally, thank you to the reviewers, whose comments helped me refine and improve my work.

NOTES

[1] On 2 October 2020, the Michigan Supreme Court ruled that Governor Whitmer did not have the power to issue or renew executive orders related to the coronavirus after 30 April 2020. The governor had extended the coronavirus state of emergency declaration on April 30th, citing the Emergency Management Act (EMA) of 1976 and the Emergency Powers of the Governor Act (EPGA) of 1945 (Alsup and Cullinane 2020). On 5 October 2020, the Michigan Department of Health and Human Services reinstated many of the governor's mandates, such as those requiring face coverings in public areas and limiting the size of gatherings, under Emergency Orders they were granted during the Spanish Flu epidemic in 1918 to protect public health during a pandemic (WXYZ 2020).

[2] This study only included and referred to surgical masks, also called "procedural masks" or "medical masks" and the N95 respirators, as both have similar value chains. The recommendations made by the OECD as a result of this study reflect the mission of this organization, one that prioritizes global economics and trade, and should be considered in light of the organization's general goals.

[3] There has been debate in academic circles about whether Facebook is really an imagined community (see Gill 2017 for arguments about why Facebook is not such a community), and while I agree with many of these arguments, I would posit that Facebook groups are indeed imagined communities, with thousands of members globally who have never met each other, do not know each other in any other context, and seemingly share the same interests or philosophies as the group's description claims. But whether Facebook or any of its groups are imagined communities or not is beyond the scope and focus of this essay.

[4] One reviewer noted that there were sewing groups that met in person. That might be the case in other places, but in Michigan, where lockdown orders were instituted early in the pandemic and were pretty strictly enforced, any group that met in person to sew would have done so in a clandestine format, its members most likely reluctant to tell others they were violating state laws, and thus I cannot report on the occurrence of this practice. The point is that most of the sewing groups were virtual as I described in the essay.

[5] Local and regional context is, of course, important here. A friend who traveled to Ann Arbor in the summer of 2020 reported people wearing masks while outside walking. Ann Arbor, as a college town, is generally considered by the rest of the state to be more progressive and forward/outward thinking than other regions.

ORCID

Sara Beth Keough ⓘ http://orcid.org/0000-0002-1710-1276

REFERENCES

Adyel, T. 2020. Accumulation of Plastic Waste during COVID-19. *Science* 369 (6509):1314–1315. September 11. doi:10.1126/science.abd9925.

Alsup, D., and S. Cullinane. 2020. Michigan Supreme Court Strikes Down Governors' Emergency COVID Powers. *CNN*, October 2. https://www.cnn.com/2020/10/02/politics/michigan-supreme-court-whitmer-covid-emergency/index.html.

Anderson, B. 1983. *Imagined Communities.* London: Verso.

Appadurai, A., edited by. 1986. *The Social Life of Things: Commodities in Cultural Perspective.* Cambridge, U.K.: Cambridge University Press.

Appadurai, A. 2012. Commodities and the Politics of Value. In *The Future as Cultural Fact: Essays on the Global Condition*, 9–60. London: Verso.

Coronavirus Cases by State. 2020. *New York Times*, 8 October. Accessed October 8, 2020. https://www.nytimes.com/interactive/2020/us/coronavirus-us-cases.html.

Gill, B. 2017. Facebook Is Not a Community. *Cyborology*, February 23. https://thesocietypages.org/cyborgology/2017/02/23/facebook-is-not-a-community/.

Guruianu, A., and N. Andrievskikh. 2019. *The Afterlife of Discarded Objects: Memory and Forgetting in a Culture of Waste.* Anderson, S.C.: Parlor Press.

Harold, C. 2020. *Things Worth Keeping: The Value of Attachment in a Material World.* Minneapolis: University of Minnesota Press.

Kaur, H., and T. Luhby. 2020. People around the Country are Sewing Masks. And Some Hospitals, Facing Dire Shortage, Welcome Them. *CNN*, March 24. https://www.cnn.com/2020/03/24/us/sewing-groups-masks-coronavirus-wellness-trnd/index.html.

Kovanis, G. 2020. New Survey Reveals How Michiganders Feel about Masks, Whitmer. *Detroit Free Press*, July 27. https://www.freep.com/story/news/local/michigan/2020/07/27/covid-survey-michigan-masks-trump/5495865002/.

Michigan State's Sparty Statue Gets Mask to Raise Coronavirus Awareness. 2020. *Lansing State Journal*, April 22. https://www.lansingstatejournal.com/picture-gallery/news/2020/04/22/michigan-states-sparty-statue-gets-mask-raise-coronavirus-awareness-east-lansing/3005021001/.

OECD. 2020. The Face Mask Global Value Chain in the COVID-19 Outbreak: Evidence and Policy Lessons. *OECD*, May 4. https://read.oecd-ilibrary.org/view/?ref=132_132616-l4ioj8cii1q&title=The-Face-Mask-Global-Value-Chain-in-the-COVID-19-Outbreak-Evidence-and-Policy-Lessons.

Oosting, J., and R. Beggin. 2020. Maskless Activists Rally against Michigan Governor Whitmer after Trump Diagnosis. *Bridge Michigan*, 2 October Accessed October 24, 2020. https://www.bridgemi.com/michigan-government/maskless-activists-rally-against-michigan-gov-whitmer-after-trump-diagnosis.

Rosalsky, G. 2020. Are High Mask Prices the Problem or the Solution? *NPR*, March 3. https://www.npr.org/sections/money/2020/03/03/811181309/are-high-mask-prices-the-problem-or-the-solution.

Slotkin, J. 2020. Protesters Swarm Michigan Capitol amid Showdown over Governor's Emergency Powers. *NPR*, May 1. https://www.npr.org/sections/coronavirus-live-updates/2020/05/01/849017021/protestors-swarm-michigan-capitol-amid-showdown-over-governors-emergency-powers.

State Health Department Reinstates Mask Mandate, Gathering Limitations in Most of Michigan. 2020. *WXYZ Detroit*, October 5. https://www.wxyz.com/news/coronavirus/state-health-department-reinstates-mask-mandate-gathering-limitations-in-most-of-michigan.

Tsing, A. 2013. Sorting Out Commodities: How Capitalist Value Is Made through Gifts. *HAU: Journal of Ethnographic Theory* 3 (1):21–43. doi:10.14318/hau3.1.003.

———. 2015. *The Mushroom at the End of the World: On the Possibility of Life in Capitalist Ruins.* Princeton, N.J.: Princeton University Press.

COVID-19 MORTALITY IN NEW YORK CITY ACROSS NEIGHBORHOODS BY RACE, ETHNICITY, AND NATIVITY STATUS

SAMANTHA FRIEDMAN⊙ and JIN-WOOK LEE

ABSTRACT. New York City has lost more lives from COVID-19 than any other American city. This study examines variation in COVID-19 deaths across neighborhoods as it relates to variation in the racial, ethnic, and nativity-status composition of neighborhoods. This topic has received little scholarly attention and is imperative to explore, given the absence of racial and ethnic specific COVID-19 mortality rates by neighborhood. New York City is a racially and ethnically segregated city, and a longstanding destination of immigrants, making some neighborhoods more susceptible to greater levels of COVID-19 mortality than others. Using ZCTA-level data on COVID-19 deaths and demographic data from the American Community Survey, our descriptive and bivariate choropleth mapping analyses reveal that a racial, ethnic, and nativity-status hierarchy exists in the geographic distribution of COVID-19 mortality. Implications of these findings are discussed as they relate to residential segregation and persistent spatial inequalities faced by neighborhoods of color.

$\mathcal{N}$ew York City (NYC) has experienced a significant share of deaths from COVID-19 within the United States (Wadhera and others 2020). At the time of this writing, four of the five counties—Bronx, Kings, Queens, and New York—that make up NYC have numbers of deaths in the top 20 counties out of the more than 3000 counties in the United States (USA Facts 2021). Out of all deaths from COVID-19 in the United States, one in 17.8 was in NYC (USA Facts 2021).

Research has shown that COVID-19 mortality rates in the United States vary by race and ethnicity (U.S. Centers for Disease Control and Prevention 2020). The percentages of Blacks and Hispanics dying from COVID-19 exceed their shares of the population, respectively, which is not the case for Whites and Asians (U.S. Centers for Disease Control and Prevention). Blacks and Hispanics in NYC also have higher rates of age-adjusted mortality than their White and Asian counterparts (Hooper and others 2020).

Racial- and ethnic-specific mortality rates are unavailable at the neighborhood level in NYC, making it difficult to pinpoint neighborhoods of color and immigrants that may be hardest hit by the pandemic. Recent research has found that nonwhite neighborhoods in NYC have higher numbers of positive COVID-19 tests than majority-White neighborhoods (DiMaggio and others 2020; Lieberman-Cribbin and others 2020). However, little research has explored the geographic variation in COVID-19 mortality rates across neighborhoods in NYC alongside the spatial variation in the racial, ethnic, and nativity-status composition of these neighborhoods. Byoungjun Kim and others (2021) and Usama Bilal and others (2021) explore variation at the neighborhood level in NYC, but they do not examine racial and ethnic groups by nativity status.

This study seeks to fill this gap and explores the geographic variation in COVID-19 mortality rates across neighborhoods in NYC, defined by Zip Code Tabulation Areas (ZCTA), as patterned by the racial, ethnic, and nativity-status compositions of the population. Moreover, we seek to advance the literature by using bivariate choropleth maps to describe the patterns of COVID-19 mortality across ZCTAs of color and immigrants in NYC. This study's use of the bivariate choropleth mapping methodology provides a systematic way for researchers to simultaneously examine the geographic variation in COVID-19 mortality rates and the racial, ethnic, and nativity-status compositions of neighborhoods in NYC, and identify neighborhoods of color and immigrants that have endured the most deaths from the pandemic.

BACKGROUND

NEW YORK CITY AS A RESIDENTIALLY SEGREGATED CONTEXT AND IMMIGRANT DESTINATION

To understand how COVID-19 mortality rates vary across NYC's neighborhoods and relate to the geographic variation in their racial, ethnic, and nativity-status composition of these neighborhoods, we first focus on two important character-istics that set NYC apart from many other cities in the United States. First, NYC has had unusually high levels of racial and ethnic residential segregation (Hotchkiss 2015; Massey and Tannen 2015). With respect to Black-White resi-dential segregation, NYC has been demarcated as a "hyper-segregated" city continuously for five decades since 1970 (Hotchkiss 2015; Massey and Tannen 2015). In 2010, the index of dissimilarity or D-score (a measure of community segregation) was 81.4 indicating that, 81.4 percent of either Blacks or Whites would have to move to achieve an even distribution within the city (Logan and Stults 2011).[1] This level of segregation falls in what is considered to be the "high range," and in 2010, this score was the second-highest score among the 200 largest cities in the United States. (Massey and Denton 1993; Logan and Stults 2011). The D-score gauging Hispanic-White residential segregation was in the high range at 65.8 and was the second-highest level of Hispanic-White segrega-tion out of the 200 largest cities in the United States (Logan and Stults 2011). The Asian-White level of segregation in NYC—50.6—was lower than the Black-White and Hispanic-White D-scores, but it was the seventh-largest value out of 200 of the largest cities in 2010 (Logan and Stults 2011).

Second, NYC has historically been a premier destination for immigrants from many countries throughout the world (Lobo and Salvo 2013). According to data from the 2019 American Community Survey (ACS), NYC had the largest foreign-born population, numbering 3.1 million, of any American city (U.S.

Census Bureau 2020a). What is unique about NYC's immigrant population is that no one group of immigrants dominates the flow of foreign-born persons to the city, resulting in large shares of each racial and ethnic group having foreign-born origins (Lobo and Salvo 2013). According to 2019 ACS data, in NYC, the share of Whites, Blacks, Hispanics, and Asians who were born outside of the United States were 22 percent, 32 percent, 40 percent, and 71 percent, respectively; in the United States, the share of Whites, Blacks, Hispanics, and Asians who were born outside of the United States were 4 percent, 10 percent, 33 percent, and 66 percent, respectively (U.S. Census Bureau 2020b). Such large shares of immigrants within each racial and ethnic group not only have an impact on how they settle in NYC, especially relative to the segregation that exists in the city, but their residential location will also have implications for how COVID-19 affects neighborhoods of color and immigrants.

RESIDENTIAL SEGREGATION AND GEOGRAPHIC VARIATION IN NEIGHBORHOOD-LEVEL HEALTH

How does residential segregation impact the health of residents in neighborhoods of color and immigrants? There is ample evidence that residential segregation is linked to adverse health outcomes and mortality among infants and adults of color, particularly for Blacks, primarily through the structural racism underlying residential segregation that constrains their residential choices (LaVeist 1993, 2003; Williams and Collins 2001; Kramer and Hogue 2009; White and Borrell 2011; Phelan and Link 2015). Residential segregation has been deemed a "fundamental cause" of racial health disparities that is independent of socioeconomic status (Williams and Collins 2001).

There are at least three mechanisms by which residential segregation affects health outcomes of people of color: through its negative impact on the socio-economic status of residents via a spatial mismatch from better-paying jobs; by its adverse impact on the social and physical environments in neighborhoods through greater levels of social disorganization and crime, poorer access to adequate supermarkets and recreational greenspaces, greater levels of exposure to environmental toxins, and worse quality housing; and through its negative impact on individual behaviors via stress like smoking, substance abuse, and a lack of physical exercise (Williams and Collins 2001; Kramer and Hogue 2009). Racial and ethnic residential segregation essentially creates unequal neighborhood and housing opportunity structures between Whites and people of color (Bernard and others 2007). Stressors, the engagement in unhealthy behaviors, and exposure to toxins are greater in segregated neighborhoods of color, and access to quality health care is lower leading to higher levels of chronic diseases and mortality for populations in these neighborhoods, relative to those in predominantly White neighborhoods (Williams and Collins 2001; Kramer and Hogue 2009).

An alternative argument about the link between segregation and health—the ethnic-density hypothesis—suggests that the concentration of people among those of their own race and ethnicity has a positive effect on their health (Bécares and others 2009; Grady and McLafferty 2007 ; Kramer and Hogue 2009; Pickett and Wilkinson 2008). The residential isolation created by discriminatory and segregating forces could result in stronger coethnic social networks, increased social support, and greater empowerment that may mitigate the harmful effects of residential segregation (LaVeist 1993; Kramer and Hogue 2009). Living among members of one's own racial and ethnic group is thought to reduce the stressors from the stigma coming from people outside of these neighborhoods, which can have a positive impact on health (Pickett and Wilkinson 2008; Bécares and others 2009). Also prominent among the mechanisms that link ethnic density to better health is the social capital of neighborhoods that can affect health by promoting informal social control, healthy norms, and increased social support (Kawachi 1999; Bécares and others 2009; Kramer and Hogue 2009).

A systematic review found mixed evidence in support of the ethnic density hypothesis (Bécares and others 2012). In the United States, for Hispanics, ethnic density appears to be positively associated with health, particularly health behaviors, but for Blacks it is more often negatively associated with health, consistent with the fundamental-causes perspective (Bécares and others 2012). Others have found more support for the ethnic density hypothesis among immigrant communities in promoting positive health outcomes (Osypuk and others 2009; McLafferty and others 2012).

SPATIAL INEQUALITY AND COVID-19 MORTALITY IN NYC NEIGHBORHOODS DURING THE PANDEMIC

The preceding discussion suggests that residential segregation puts predominantly Black neighborhoods at a unique disadvantage in terms of health outcomes, relative to neighborhoods that are disproportionately Hispanic and Asian and predominated by immigrants. Against this backdrop, we now must consider the spread of COVID-19, a highly contagious virus, and the geographic distribution of COVID-19 mortality. Ethnically dense neighborhoods may not be shielded from the negative consequences of COVID-19 as is the case from chronic health conditions. Residential isolation among coethnics and household crowding, which both foster social networks and support, can create significant vulnerability to populations when it comes to highly contagious diseases (Acevedo-Garcia 2000; 2001).

In NYC, Blacks and Hispanics will already be at risk of greater exposure to COVID-19 because of the underlying poorer health present in largely Black and Hispanic neighborhoods in terms of diabetes, hypertension, and cardiovascular disease resulting from segregation (Williams and Collins 2001). These chronic conditions are related to the vulnerability of the Black and Hispanic population to COVID-19 (Ssentongo and others 2020; Bajgain and others 2021). Areas with large shares of Asians and Hispanics will also be at risk of greater levels of COVID-

19 mortality because of their higher levels of household crowding than in predominately White neighborhoods (Rosenbaum and Friedman 2007).

Sara McLafferty (2010) discusses the idea that geographic mobility is an important phenomenon that could raise or lower population exposure to infectious pathogens. This is particularly important in NYC. As mentioned above, one of the consequences of residential segregation that results from structural racism in the housing market against Blacks is that there is a spatial mismatch between where they live and work (Kain 2004). Daily, Blacks are likely to be highly mobile throughout NYC and potentially have a greater exposure to COVID-19 than other groups who do not experience this spatial mismatch. One study finds that the nature of the commuting patterns of Blacks that results from spatial mismatch and the segregated nature of Black residential areas produces high incidence rates of COVID-19 among African-Americans (Bassolas and others 2021).

Another way that mobility may affect the exposure of populations in neighborhoods that are composed largely of nonwhites and immigrants to COVID-19 is through employment in transportation occupations that involve contact with people in transit (for example, bus and taxi drivers). In their study of variation in COVID-19 positive tests across ZCTAs in NYC, Milena Almagro and Angelo Orane-Hutchinson (2020) find that the percentage employed in transportation occupations is positively associated with the percentage of COVID-19 positive tests. In NYC in 2019, among males, 3.5 percent of Whites were employed in transportation occupations as compared to 10.5 percent of Blacks, 10.0 percent of Hispanics, and 11.9 percent of Asians (U.S. Census Bureau 2020d). In NYC, in transportation, warehousing, and utilities occupations, immigrants made up 53 percent of the workforce (DiNapoli and Bleiwas 2019). It is likely that COVID-19 mortality will be greater in nonwhite and immigrant neighborhoods through the exposure to the virus via these occupations and because of the segregated nature of Black, Hispanic, and Asian neighborhoods in NYC, thereby concentrating contact with coethnics and magnifying exposure of the virus to these populations.

The final way mobility could be linked to COVID-19 exposure is via the out-migration that occurred from NYC during the pandemic, which could have lowered exposure to the virus. According to Kevin Quealy (2020), the richest neighborhoods in NYC, which overlap with the Whitest neighborhoods, experienced the most significant exodus from NYC as of 1 May 2020. This out-migration could have made the COVID-19 mortality rate lower in predominantly White neighborhoods than in predominantly nonwhite neighborhoods. However, analyses by Kim and others (2021) that use mobility data from Quealy (2020) does not find a significant association between out-migration and COVID-19 mortality rates.

This suggests that there is likely to be a racial and ethnic hierarchy with respect to the way COVID-19 mortality rates are distributed across neighborhoods by racial, ethnic, and nativity status composition. We expect that COVID-19 mortality rates will be the highest in neighborhoods with large shares of native-born Black population and the lowest in neighborhoods with large shares of native-born White population, where many people have left NYC. Neighborhoods with large shares of foreign-born Black and native- and foreign-born Hispanic populations will likely fall in between these two extremes because these populations have experienced constraints in their residential opportunities but not to the extent, historically, that native-born Blacks have faced. However, it is unclear where neighborhoods with large percentages of Asians and foreign-born Whites will fall on this continuum. The contagious nature of the virus, high levels of ethnic density, and nature of their employment will likely put them more at risk than neighborhoods that are largely native-born White, but the benefits of ethnic density do not put them as much at risk as predominately Black and Hispanic neighborhoods.

DATA AND METHODS

Two main data sources are used for our analysis of COVID-19 mortality rates in NYC neighborhoods. Data on COVID-19 deaths at the ZCTA level are acquired from data released daily by the New York City Department of Health, which we obtained on 1 February 2021 (NYC DOH [New York City Department of Health] 2021). Data on the racial, ethnic, and nativity-status composition of ZCTAs are acquired from the 2014–2018, five-year release of the American Community Survey (ACS) available via the IPUMS NHGIS website maintained by the University of Minnesota Population Center (https://www.nhgis.org/). Our unit of analysis is at the neighborhood level, defined by ZCTAs. In total, there are 177 ZCTAs included in our analysis. While ZCTAs are conveniently used by researchers as a proxy for neighborhoods, they are not without their limitations. ZCTAs are areas that align with postal-service distribution and are not defined as meaningful communities by the residents themselves. Research has shown that statistical relationships are often more consistent at the census-tract and block-group levels than at the ZCTA level because of more heterogeneity found within ZCTAs, although with respect to studies focused on mortality, the evidence is not substantially different by geographic level (Krieger and others 2002).

The primary outcome of interest is the COVID-19 mortality rate for each ZCTA, calculated as the total number of COVID-19 deaths in a ZCTA per 100,000 population. Our main independent variable of interest gauges the racial, ethnic, and nativity status composition of ZCTAs. We measure this composition with the following variables: the percentages of native- and foreign-born Whites, Blacks, Hispanics, and Asians. Because we are examining the native- and foreign-born segments within racial and ethnic groups, we are limited to the ACS summary tables provided by the U.S. Census Bureau. They provide

native- and foreign-born segments only within the non-Hispanic White population and do not disaggregate other racial groups by their Hispanic origin.[2]
Thus, in our data, Blacks include Hispanics and non-Hispanics, and our Black
and Hispanic compositional variables are not mutually exclusive. However, in
NYC, Blacks and Hispanics are, for the most part, distinct groups. In 2019, only
12 percent of Blacks identified that they were of Hispanic origin, and among
Hispanics, only 10 percent identified as being Black (U.S. Census Bureau 2020d).

To examine the association between racial, ethnic, and nativity-status composition and COVID-19 mortality, first we present descriptive statistics. We run
bivariate correlations between COVID-19 mortality rates and each racial, ethnic,
nativity-status compositional variable, separately. In addition, we calculate
weighted means of COVID-19 deaths per 100,000 for each racial, ethnic, nativity-
status compositional group. To do this, we take the COVID-19 death rate in each
ZCTA and multiply it by the size of the group of interest, sum these values, and
divide by the total number of that group across all ZCTAs (Malega and Stallings
2016).

Then, we show a univariate choropleth map of the spatial variation of COVID-
19 deaths per 100,000 population across ZCTAs in NYC.[3] Finally, we present
bivariate choropleth maps of variation in COVID-19 mortality across ZCTAs by
their racial, ethnic, and nativity status composition. Bivariate choropleth mapping is an underutilized technique, although previous research recommends it
(Leonowicz 2006; Biesecker and others 2020). For our purposes, this mapping
technique is ideal because two ZCTA-level variables can be mapped simultaneously. Because COVID-19 mortality rates are not available by race, ethnicity, or
nativity status, the bivariate choropleth mapping technique shows areas that
have a range in COVID-19 mortality rates simultaneously by the racial, ethnic,
and nativity-status composition of areas. Thus, this mapping procedure reveals
whether areas with disproportionately high shares of specific racial, ethnic, and
nativity-status groups are also areas with high COVID-19 mortality rates. We adopt
the color schemes recommended by Joshua Stevens (2015) because it is easier for
readers to distinguish between areas with low and high COVID-19 mortality rates
and low and high native- and foreign-born shares of specific racial and ethnic
groups. Stevens (2015) offers four different color templates that are ideal. We use
one for each of our four racial and ethnic groups. In all the maps, we keep the
low-mid-high color ramp for the COVID-19 mortality rate variable in the same
color scheme.

For the categorization of the variables in our maps, we use the quantile
classification method (ArcGIS Pro 2021). Specifically, we split the values of each
variable into three, equally sized ranges. In the bivariate choropleth maps, this
results in nine categories of colors that are visualized on the 3 × 3 cross-
tabulation legend. This method is preferable for making comparisons across
several maps.

RESULTS

DESCRIPTIVE STATISTICS

Column 1 of Table 1 presents the bivariate correlation coefficients. The first result, for native-born, non-Hispanic Whites[4] indicates that there is a negative and statistically significant, moderate correlation between percentage native-born Whites and COVID-19 mortality rates across ZCTAs in NYC. As the percentage of native-born Whites increases, the COVID-19 mortality rate decreases. While there is also a negative correlation between the percentage of foreign-born Whites and COVID-19 mortality, this relationship is not significant. Both the percentages of native-born and foreign-born Blacks are significantly and positively associated with COVID-19 mortality rates in ZCTAs. The magnitude of the association is stronger between the percentage of native-born Blacks and COVID-19 mortality rates than it is between the percentage of foreign-born Blacks and COVID-19 mortality rates. Both the percentages of native-born and foreign-born Hispanics are also significantly and positively associated with COVID-19 mortality rates in ZCTAs at similar magnitudes. With respect to the percentages of native- and foreign-born Asians, however, there are negative associations with COVID-19 mortality rates, although the latter association is not statistically significant.

Column 2 of Table 1 reports the weighted means of COVID-19 mortality rates by race, ethnicity, and nativity-status group. Consistent with the results in column 1, native-born Whites live in ZCTAs, on average, with the lowest level of COVID-19 mortality rate, 213.34 per 100,000 population. Native- and foreign-born Blacks, on the other hand, live in ZCTAs with the highest averages, 291.7 and 301.08 per 100,000 population, respectively. Native- and foreign-born Hispanics live in ZCTAs with COVID-19 mortality rates in the middle, and native-

TABLE 1—DESCRIPTIVE ANALYSES OF ASSOCIATION BETWEEN RACIAL, ETHNIC, AND NATIVE-STATUS COMPOSITION AND COVID-19 MORTALITY RATES ACROSS ZCTAS, NEW YORK CITY

GROUP	BIVARIATE CORRELATION COEFFICIENTS (1)	MEAN COVID-19 MORTALITY RATE (WEIGHTED BY GROUP POPULATION SIZE) (2)
Percentage:		
Native-born White	−0.522***	213.34
Foreign-born White	−0.112	254.22
Native-born Black	0.338***	291.70
Foreign-born Black	0.279***	301.08
Native-born Hispanic	0.290***	277.85
Foreign-born Hispanic	0.276***	288.73
Native-born Asian	−0.220**	249.02
Foreign-born Asian	−0.033	267.91
N	177	177

and foreign-born Asians live in ZCTAs with COVID-19 mortality rates that are slightly lower and slightly higher, respectively, relative to those in ZCTAs where foreign-born Whites reside.

The results in Table 1 suggest that the racial, ethnic, and nativity status composition of ZCTAs is associated with the variation in COVID-19 mortality rates and that significant variation exists in COVID-19 mortality rates depending upon the racial, ethnic, and nativity-status composition that is present in ZCTAs. We now turn to the mapping analysis to explore this variation further. First, we present the univariate choropleth map of COVID-19 mortality rates. Then, we present bivariate choropleth maps for native-born Whites and Blacks and foreign-born Hispanics and Asians. Our bivariate analysis in Table 1 shows that the percentage of these groups within ZCTAs is significantly associated with variation in ZCTA-level COVID-19 mortality rates, except for foreign-born Asians. Therefore, we wish to explore the variation in greater depth through our maps.[5]

COVID-19 MORTALITY MAP

Areas on the map in Figure 1 in the darkest aqua color indicate ZCTAs with the highest COVID-19 mortality rates. The share of ZCTAs in each borough that fall into the highest COVID-19 mortality rate is 56 percent in the Bronx, 41 percent in the Queens, 33 percent in Staten Island, 30 percent in Brooklyn, and only 14 percent in Manhattan. In the Bronx, places like Kingsbridge-Riverdale, Hunts Point-Mott Haven, Pelham-Throgs Neck, and northeast Bronx have the highest levels of COVID-19 mortality rates. In Manhattan, the two areas most affected by high levels of COVID-19 mortality are in Washington Heights-Inwood and East Harlem. In Queens, several ZCTAs in west Queens, Ridgewood-Forest Hills, southwest Queens, Jamaica, and Rockaway exhibit levels of COVID-19 mortality rates in the highest category. In Brooklyn, ZCTAs in Coney Island-Sheepshead Bay, Borough Park, east Flatbush-Flatbush, Bedford Stuyvesant-Crown Heights, and East New York have the highest levels of COVID-19 mortality rates. Finally, in Staten Island, ZCTAs in Willowbrook, South Beach-Tottenville, and Stapleton-St. George display the highest levels of COVID-19 mortality rates.

WHITE NATIVE-BORN MAP

Figure 2 presents the first of our bivariate choropleth maps. It examines COVID-19 mortality rates by the percentage native-born White across ZCTAs in NYC. Areas that are shaded the dark aqua color at the top left of the nine-square legend are in the highest category of COVID-19 mortality rates, but in the lowest category of percentage of native-born Whites. We focus our attention on the two categories in the upper-right portion of the nine-square legend that are shaded dark brown and brown. The dark brown category shows ZCTAs with the highest level of COVID-19 mortality rates and the highest category of the percentage of

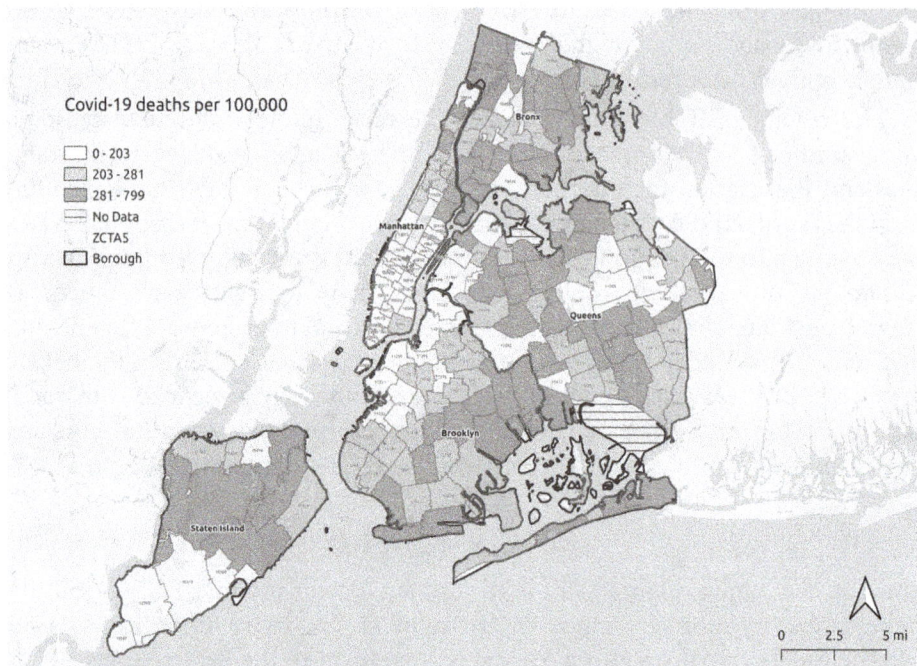

FIG. 1—COVID-19 mortality rates by ZCTAs, New York City.

native-born Whites. The brown category indicates ZCTAs with levels of COVID-19 mortality rates in the middle category and the highest category of the percentage of native-born Whites. As we discuss all the bivariate choropleth maps, we will focus on these two categories in the upper-right portion of the nine-square legend because they will reveal the intersection between middle-to-high levels of COVID-19 mortality rates and the highest level of the racial, ethnic, and nativity-status compositional group of interest. To make it easy for the reader to identify these categories on the map, we have outlined them in black. We also focus on the category on the bottom-right side of the nine-square legend, which shows ZCTAs with lowest levels of COVID-19 mortality rates and the highest category of the racial, ethnic, and nativity-status group of interest. We are interested in whether ZCTAs with shares of each group in the highest category reside in places with the lowest levels of COVID-19 mortality rates.

Just eight out of the 177 ZCTAs in NYC fall into the dark brown category—a neighborhood in Kingsbridge-Riverdale, Bronx; a neighborhood in Borough Park in Brooklyn; ZCTAs in Flushing-Clearview, Ridgewood-Forest Hills, southwest Queens, and Rockaway in Queens; and in Willowbrook and south Beach-Tottenville of Staten Island. Relative to Figure 1, the majority of ZCTAs where COVID-19 mortality rates are high are areas with lower shares of native-born White

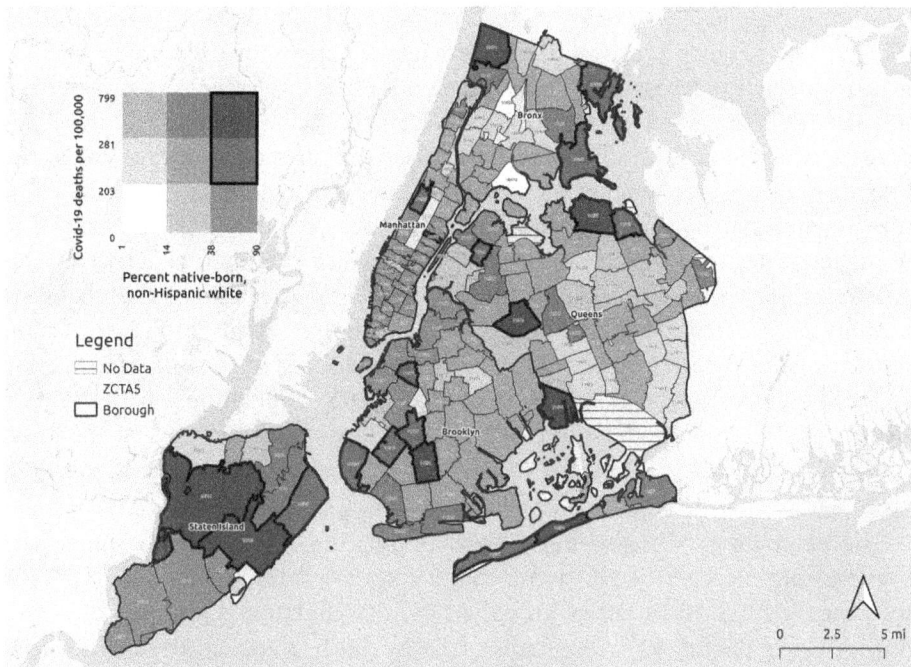

Fig. 2—COVID-19 mortality rate and percent native-born, non-Hispanic Whites by ZCTAs, New York City.

populations and greater shares of nonwhite populations. Regarding the middle level of COVID-19 mortality among ZCTAs with the highest level of the share of native-born White population, only 13 ZCTAs fall into this category. These ZCTAs are dispersed throughout the five boroughs of NYC.

Most of the neighborhoods in the highest category of the share of native-born White population—64 percent of ZCTAs in this category—fall in the red category at the lower right-hand corner of the legend. These ZCTAS, which number 38, are areas with the lowest levels of COVID-19 mortality rates. For example, most of the middle and southern sections of Manhattan fall in this category, as well as the southern portion of Staten Island; neighborhoods in Greenpoint and Downtown-Heights-Park Slope, Brooklyn; and in Long Island City-Astoria and Bayside-Little Neck, Queens. Notably, most of the ZCTAs in the Bronx, Brooklyn, and Queens fall in the two blue categories at the upper- and middle-leftmost corner of the legend. These neighborhoods have the lowest shares of native-born White population and COVID-19 mortality rates that are either in the highest or second highest categories. Thus, it is clear from the map that most neighborhoods with the highest level of percentage of native-born White population have the lowest levels of deaths from COVID-19 in NYC.

BLACK NATIVE-BORN MAP

The next map, Figure 3, shows the percentage of native-born Blacks by levels of COVID-19 mortality across NYC neighborhoods. It is profoundly different from the previous map. Most of the ZCTAs with percentages of native-born Black population in the highest range fall into either the highest or middle categories of COVID-19 deaths per 100,000 population. There is a total of 54 ZCTAs in these mid-to-high COVID-19 mortality rate categories out of the 59 ZCTAs that fall in the highest category of percentage native-born Black population. These ZCTAs tend to overlap with the ZCTAs shown in the middle and highest COVID-19 mortality rate category in the univariate choropleth map in Figure 1. One difference from Figure 1 is the absence of neighborhoods in the northern portion of Queens. Only five neighborhoods contain Black populations in the highest category and the lowest levels of COVID-19 mortality, shaded in pink in the lower right-hand corner of the legend, which is in sharp contrast to the map of the percentage of native-born Whites in Figure 2.

The percentage of native-born Black population in the highest range and middle-to-high levels of COVID-19 mortality rates are found mostly in ZCTAs in the Bronx, central and eastern Brooklyn, and southeastern Queens. The distribution of neighborhoods with shares of native-born Blacks in the highest level and mid-to-high rates of COVID-19 mortality is widespread in the Bronx, and

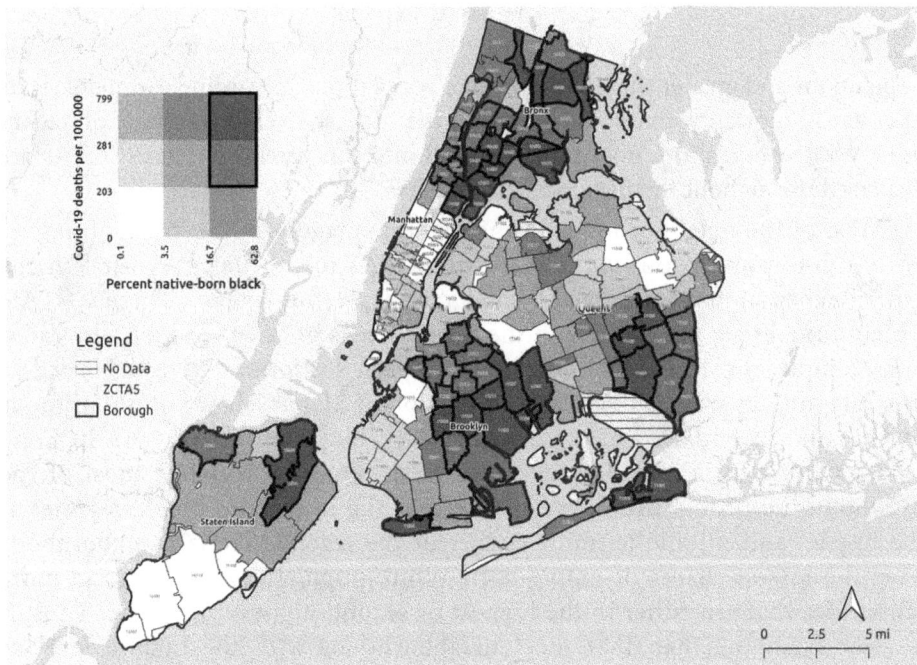

FIG. 3—COVID-19 mortality rate and percent native-born Blacks by ZCTAs, New York City.

includes ZCTAs in northeast Bronx, Fordham-Bronx Park, Crotona-Tremont, High Bridge-Morrisania, Hunts Point-Mott Haven, and Pelham-Throgs Neck; in central and eastern Brooklyn, ZCTAs are found in Bedford Stuyvesant-Crown Heights, Williamsburg-Bushwick, Flatbush, Canarsie-Flatlands, and East New York; and in southeastern Queens, the neighborhoods are in Jamaica, southeast Queens, and the eastern portion of Rockaway. Many of these neighborhoods exhibit higher levels of poverty, unemployment, and crime, than found in the city (NYU Furman Center 2018). In Manhattan, few neighborhoods are in the highest category of the share of native-born Blacks and in the middle-to-highest ranges of COVID-19 deaths per 100,000 population; and they are in East Harlem, central Harlem-Morningside Heights, and Washington Heights-Inwood. In Staten Island, ZCTAs in Port Richmond and Stapleton-St. George fall into this range.

HISPANIC FOREIGN-BORN MAP

Figure 4 presents a bivariate choropleth map of COVID-19 mortality rates by the percentage of the population that is foreign-born Hispanic. Like the previous map for the native-born Black population, it is evident that most of the ZCTAs in the highest category of percentage of foreign-born Hispanic populations fall into the middle or highest categories of COVID-19 death rates, which are denoted by the two shades of darker green in the upper-right corner and far-right middle section of the legend. Out of the 59 ZCTAs that fall in the highest category of the percentage of foreign-born Hispanic population, 49 or 83 percent fall into the middle or highest categories of COVID-19 mortality rates, which is slightly lower than was the case for native-born Blacks in Figure 3. These 49 ZCTAs tend to overlap with the middle and highest COVID-19 mortality rate categories in the univariate choropleth map of COVID-19 mortality in Figure 1. The main difference, however, is that Figure 4 contains fewer neighborhoods in Brooklyn. In contrast to Figure 2 of the native-born White population, only 10 ZCTAs in Figure 4 are in the highest category of percentage of foreign-born Hispanic population and the lowest levels of COVID-19 mortality, which is shaded the lightest shade of green in the lower right-hand corner of the legend. However, the share in this category, 17 percent, is double that found in Figure 3 for the percentage native-born Black.

There are similarities but also distinct differences between Figures 3 and 4, in terms of the distribution of neighborhoods where foreign-born Hispanic population is the highest and COVID-19 mortality rates are in the middle and highest categories, relative to areas with the highest levels of native-born Black population in the same COVID-19 mortality categories. For example, there is overlap in the distributions of foreign-born Hispanic and native-born Black populations in these middle-to-high COVID-19 mortality rate ranges in ZCTAs in the central and southern sections of the Bronx in neighborhoods like Crotona-Tremont, High Bridge-Morrisania, and Hunts Point-Mott Haven; and in northern Manhattan in

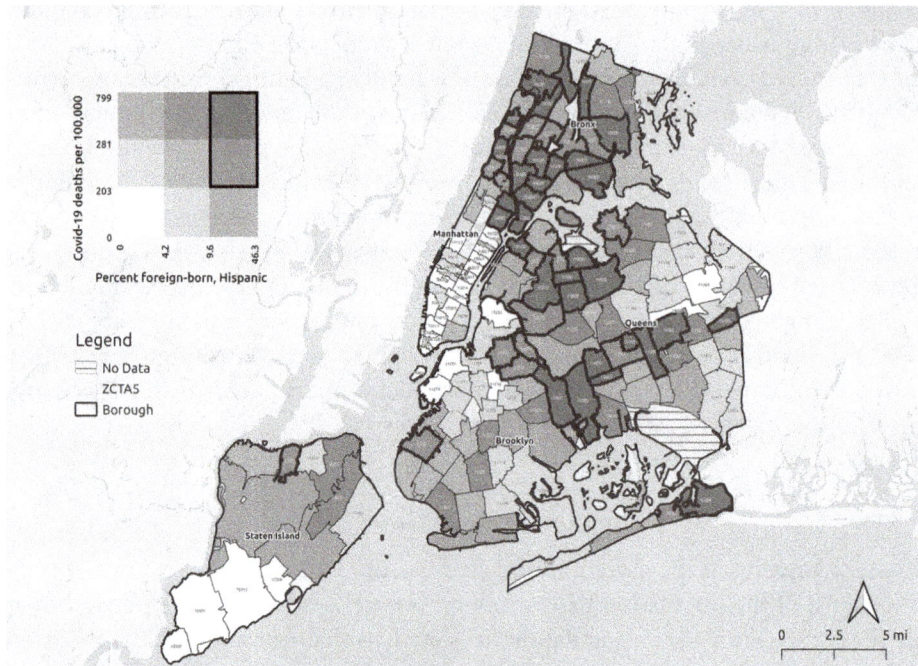

FIG. 4—COVID-19 mortality rate and percent foreign-born Hispanics by ZCTAs, New York City.

Washington Heights-Inwood, central Harlem-Morningside Heights, and East Harlem.

In Brooklyn and Queens, however, Figure 4 shows that the neighborhoods in the highest category of percentage foreign-born Hispanics and that have middle-to-high levels of COVID-19 deaths are different in their distribution than those in the highest category of the share of native-born Black population and in the same COVID-19 mortality categories. In Brooklyn, there are very few neighborhoods that are in the highest category of percentage of foreign-born Hispanic population and middle-to-high levels of COVID-19 deaths, and they include ZCTAs in Sunset Park, Williamsburg-Bushwick, and East New York. In Queens, the differences in the distributions of foreign-born Hispanic population and native-born Black population are also evident. ZCTAs in the highest category of percentage of foreign-born Hispanic population are more widespread throughout the northeastern and central sections of Queens in Long Island City-Astoria, west Queens, Flushing-Clearview, and southwest Queens. As was the case for neighborhoods that fall in the highest category of the share of native-born Black populations and have high levels of COVID-19 mortality rates, many of the neighborhoods that are in the highest level of the share of foreign-born

Hispanic population and in the highest category of COVID-19 mortality rates exhibit higher levels of poverty, unemployment, and crime, than found in the city (NYU Furman Center 2018).

ASIAN FOREIGN-BORN MAP

Figure 5 presents the choropleth bivariate map of COVID-19 mortality rates by the percentage of the population that is foreign-born Asian. It is evident that many ZCTAs in the highest category of the percentage of foreign-born Asian populations fall into the middle or highest categories of COVID-19 deaths (that is, 35 out of the 59 ZCTAs or 59 percent in the highest category of foreign-born Asian population), which are denoted by the two shades of brown in the upper-right corner and far-right middle section of the legend. However, this level is far lower than was the case in the native-born Black and foreign-born Hispanic maps in Figures 3 and 4, respectively. There is some overlap of these 35 ZCTAs with the middle to highest COVID-19 mortality rate categories in the univariate choropleth map of COVID-19 mortality shown in Figure 1, but not as much as was evident between the maps of native-born Blacks and foreign-born Hispanics. Most of the ZCTAs in the highest category of the share of foreign-born Asian population and highest levels of COVID-19 mortality rates are in Queens; only a few are in

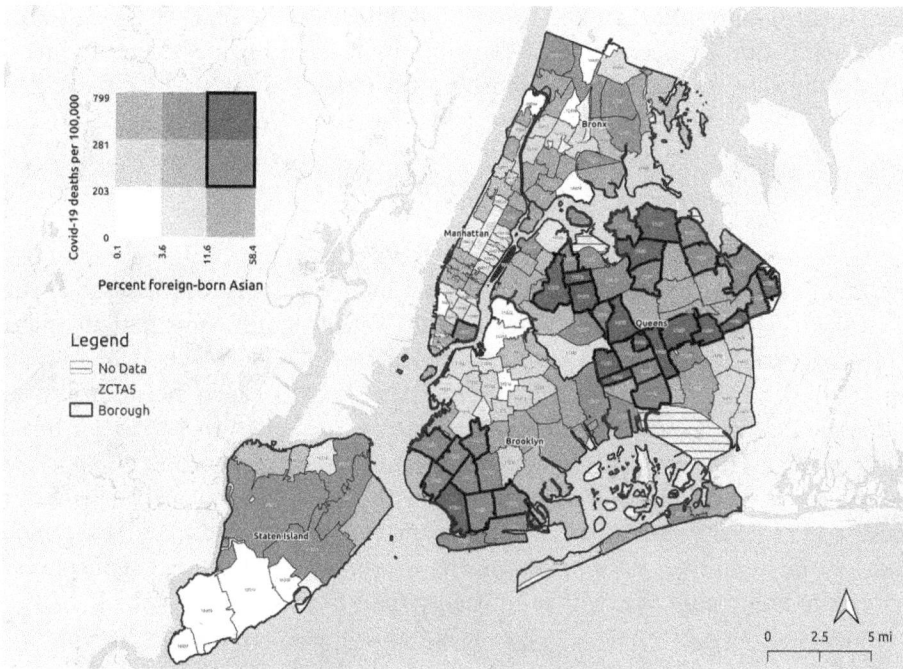

FIG. 5—COVID-19 mortality rate and percent foreign-born Asians by ZCTAs, New York City.

Brooklyn. Absent from the foreign-born Asian map are ZCTAs in the highest COVID-19 mortality category in the Bronx, Manhattan, and Staten Island.

In Figure 5, there are many ZCTAs—24 or 41 percent—that fall in the lowest level of COVID-19 mortality rates and the highest level of the share of foreign-born Asian population, which is shaded orange in the lower right-hand corner of the legend. Relative to the maps of the other racial, ethnic, and nativity-status compositional groups, there is a more even split of neighborhoods that are in the highest category of the share of foreign-born Asian population and fall into the lowest, middle, and highest levels of COVID-19 deaths per 100,000 population than is the case when examining ZCTAs that are in the highest levels of the shares of native-born Black and foreign-born Hispanic populations. However, relative to Figure 2, the share of ZCTAs with the highest level of the share of foreign-born Asians and the lowest COVID-19 mortality rates is lower than that for native-born Whites in the same COVID-19 mortality rate category (41 percent versus 64 percent).

The distribution of neighborhoods where foreign-born Asian population is in the highest category and COVID-19 mortality rates are in the middle-to-high categories is distinct from the previous maps, although there are some similarities in the ZCTAs in Figure 4 focused on the percentage of foreign-born Hispanics. Starting with the similarities, there is overlap occurring in Queens in some ZCTAs in southwest Queens, Jamaica, west Queens and one ZCTA in Flushing-Clearview. The distinct pattern of the intersection between middle-to-high COVID-19 mortality rates and in the highest level of the share of foreign-born Asian population is found in ZCTAs primarily located in northeastern Queens in the following neighborhoods: Flushing-Clearview, Bayside-Little Neck, and southeast Queens; and unsurprisingly in Manhattan in Union Square-Lower East Side.

DISCUSSION AND CONCLUSIONS

The main objective of this study was to examine the association between spatial variation in COVID-19 mortality rates and the racial, ethnic, and nativity status composition of populations across neighborhoods in NYC. We sought to identify neighborhoods of color and immigrants in NYC that have been hardest hit by the pandemic, using bivariate correlations and weighted means and a novel, bivariate choropleth mapping approach. The bivariate choropleth approach allowed us to overcome the data limitation of not having available neighborhood-level mortality rates disaggregated by race, ethnicity, and nativity status. In addition, we attempted to offer insights as to why some of these neighborhoods were more vulnerable to COVID-19 mortality than others.

Our descriptive analyses and bivariate choropleth maps reveal that a hierarchy emerges based on racial, ethnic, and, to a lesser extent, nativity-status composition, in the neighborhoods hardest hit by COVID-19 mortality.

Neighborhoods in the highest category of the share of native-born White population have fared best in terms of experiencing the lowest levels of COVID-19 mortality rates, and neighborhoods in the highest category of the shares of native- and foreign-born Black populations have fared the worst. The average COVID-19 mortality rate among foreign-born Whites is greater than that for native-born Whites, but the native-born Asian rate is lower than that for foreign-born Whites. Neighborhoods in the highest category of the percentages of native- and foreign-born Hispanic populations fare somewhat better than neighborhoods in the highest category of the share of Blacks, but the difference is not that profound. On the other hand, neighborhoods in the highest category of the share of foreign-born Asians fall toward the middle of the hierarchy of places on the continuum of COVID-19 mortality rates.

What is clear from our analysis is that the residential segregation of Blacks, Hispanics, and Asians from Whites makes neighborhoods with large nonwhite shares of the population more vulnerable to COVID-19 mortality, although at varying degrees. But the underlying reasons for the vulnerability to COVID-19 differs by the racial, ethnic, and nativity-status component of the population. Consistent with the fundamental-causes perspective, decades of structural racism and residential segregation make predominantly native-born Black neighborhoods more vulnerable to experiencing concentrated disadvantage, poorer health, and increased levels of mortality, which puts them at greater risk of mortality from COVID-19 (LaVeist 1993, 2003; Williams and Collins 2001; Kramer and Hogue 2009; White and Borrell 2011; Phelan and Link 2015). Moreover, the spatial mismatch that results from residential segregation increases the vulnerability of the populations in these neighborhoods, because of their daily travel outside of their neighborhoods to go to work (Kain 2004). Taken together, these factors explain why the largest number of predominantly native-born Black neighborhoods experienced the highest levels of COVID-19 mortality.

Our analyses, however, show that neighborhoods in the highest category of the shares of foreign-born Hispanics and Asians are also vulnerable to COVID-19 mortality. The fact that their vulnerability is not as high as that of neighborhoods falling in the highest category of percentage of native-born Blacks is suggestive of some evidence for the ethnic-density hypothesis (Kawachi 1999; Pickett and Wilkinson 2008; Bécares and others 2009; Kramer and Hogue 2009). Particularly in the case of neighborhoods falling in the highest category of the percentage of foreign-born Asians, it is likely that the ethnic density has a positive effect on the health of these areas. In addition, at least half of native-born Asians are under 18 years of age in NYC and likely live with foreign-born Asian parents (Friedman and others 2021). Because younger people have much lower levels of risk of dying from COVID-19, this demographic distribution could be driving down the overall COVID-19 mortality rates, even in neighborhoods with greater shares of foreign-born Asians. The vulnerability to COVID-19 mortality in neighborhoods with larger shares of immigrant Asians and Hispanics likely stems from the participation in transportation occupations of populations in these areas that exposes them to the virus (DiNapoli and Bleiwas 2019; Almagro and Orane-Hutchinson 2020; U. S. Census Bureau 2020c).

Neighborhoods with the lowest levels of COVID-19 mortality were areas that had the highest levels of the percentage of native-born Whites. This likely stems from the benefits accrued to Whiter neighborhoods via racial and ethnic segregation in terms of the health of Whites in these areas and the fact that amenities such as jobs, access to the best health care, supermarkets, and recreational greenspaces make it easy for residents in these areas to maintain healthy lifestyles and significantly lowers their levels of chronic health conditions (Williams and Collins 2001). Moreover, during the pandemic residents in largely White neighborhoods were the most likely to flee NYC, thereby potentially lowering the COVID-19 mortality rates in these areas (Quealy 2020).

The analyses presented here makes clear that the pandemic has significantly exacerbated spatial inequalities in NYC that already existed well before COVID-19. Decades of racial and ethnic residential segregation and disinvestment, and the resultant poverty and unemployment, have contributed to COVID-19 mortality, particularly among Blacks, but also Hispanics. To put the Black-White racial inequality in perspective, Elizabeth Wrigley-Field (2020: 21854) finds that "for White life expectancy in 2020 to fall to the level of the best-recorded Black life expectancy would require an estimated 700,000 to 1 million excess White deaths." The results in this study should help guide policy makers to invest more resources in neighborhoods of color in NYC and particularly in neighborhoods with larger shares of native-born Blacks, in terms of housing, jobs, and health care to improve the population health in these neighborhoods.

ACKNOWLEDGMENTS

We would like to thank Dr. Tabassum Insaf, Assistant Professor of the Department of Epidemiology and Biostatistics, and Dr. Temilayo Adeyeye, Assistant Research Professor of the Department of Environmental Health Sciences, both from the University at Albany, SUNY, for their research assistance and technical advice. We would also like to thank the New York State COVID-19 Minority Health Disparities Team at the University at Albany, SUNY for the valuable input on our presentation of an earlier version of this research. Finally, we acknowledge the very helpful input of the anonymous reviewers that significantly improved this paper.

FUNDING

Support for this research was provided by the Center for Social and Demographic Analysis at the University at Albany, SUNY.

NOTES

[1] Logan and Stults (2011) calculated D-scores based upon tract-level data. D-scores are, however, sensitive to geographic scale, and those based upon larger levels of geography (e.g., tracts versus block groups) tend to be lower in value than D-scores based upon smaller levels of

geography (Wong 1997). Notably, New York City has high D-scores even using tracts as the unit of analysis (Massey and Tannen 2015).

[2] A special tabulation from the U.S. Census Bureau is needed to obtain these data. We cannot aggregate PUMS data because the ZCTA geography is unavailable within the publicly-available PUMS data.

[3] For the projection of the maps, we use the UTM standard by New York State (https://gis.ny.gov/coordinationprogram/workgroups/wg_1/related/standards/datum.htm). All publications by the New York State Department of Health and other state agencies use this projection.

[4] Hereafter, for simplicity, we just refer to non-Hispanic Whites as Whites.

[5] We limit our analyses to these groups because multivariate analyses in Friedman and others (2021) reveal an association between the percentages of these racial, ethnic, and nativity-status groups and COVID-19 mortality at the ZCTA-level. The maps for foreign-born Whites and Blacks and native-born Hispanics and Asians are available upon request of the authors.

ORCID

Samantha Friedman ⓘ http://orcid.org/0000-0001-8412-9618

REFERENCES

Acevedo-Garcia, D. 2000. Residential Segregation and the Epidemiology of Infectious Diseases. *Social Science & Medicine* 51 (8):1143–1161. doi:10.1016/S0277-9536(00)00016-2.

_____. 2001. Zip Code-Level Risk Factors for Tuberculosis: Neighborhood Environment and Residential Segregation in New Jersey, 1985-1992. *American Journal of Public Health* 91 (5):734.

ArcGIS Pro. 2021. Data Classification Methods – Quantile. https://pro.arcgis.com/en/pro-app/latest/help/mapping/layer-properties/data-classification-methods.htm.

Almagro, M., and A. Orane-Hutchinson. 2020. JUE Insight: The Determinants of the Differential Exposure to COVID-19 in New York City and Their Evolution over Time. *Journal of Urban Economics*: 103293. doi:10.1016/j.jue.2020.103293.

Bajgain, K. T., S. Badal, B. B. Bajgain, and M. J. Santana. 2021. Prevalence of Comorbidities among Individuals with COVID-19: A Rapid Review of the Current Literature. *American Journal of Infection Control* 49 (2):238–246. doi:10.1016/j.ajic.2020.06.213.

Bassolas, A., S. Sousa, and V. Nicosia. 2021. Diffusion Segregation and the Disproportionate Incidence of COVID-19 in African American Communities. *Journal of the Royal Society Interface* 18 (174):20200961. doi:10.1098/rsif.2020.0961.

Bécares, L., J. Nazroo, and M. Stafford. 2009. The Buffering Effects of Ethnic Density on Experienced Racism and Health. *Health & Place* 15 (3):700–708. doi:10.1016/j.healthplace.2008.10.008.

Bécares, L., R. Shaw, J. Nazroo, M. Stafford, C. Albor, K. Atkin, K. Kiernan, R. Wilkinson, and K. Pickett. 2012. Ethnic Density Effects on Physical Morbidity, Mortality, and Health Behaviors: A Systematic Review of the Literature. *American Journal of Public Health* 102: e33_e66. doi:10.2105/AJPH.2012.300832.

Bernard, P., R. Charafeddine, K. L. Frohlich, M. Daniel, Y. Kestens, and L. Potvin. 2007. Health Inequalities and Place: A Theoretical Conception of Neighborhood. *Social Science & Medicine* 65:1839–1852. doi:10.1016/j.socscimed.2007.05.037.

Biesecker, C., W. E. Zahnd, H. M. Brandt, S. A. Adams, and J. M. Eberth. 2020. A Bivariate Mapping Tutorial for Cancer Control Resource Allocation Decisions and Interventions. *Preventing Chronic Disease: Public Health, Practice, and Policy* 17 (E01):1–9. doi:10.5888/pcd17.190254.

Bilal, U., L. P. Tabb, S. Barber, and A. V. Diez Rouz. 2021. Spatial Inequities in COVID-19 Testing, Positivity, Confirmed Cases, and Mortality in 3 U.S. Cities. *Annals of Internal Medicine*. March 30, 2021. doi:10.7326/M20-3936.

DiMaggio, C., M. Klein, C. Berry, and S. Frangos. 2020. Black/African American Communities are at Highest Risk of COVID-19: Spatial Modeling of New York City Zip Code-Level Testing Results. *Annals of Epidemiology* 51: 7–13. doi:10.1016/j.annepidem.2020.08.012.

DiNapoli, T. P., and K. B. Bleiwas. 2019. New York City Employment Trends. *New York State Comptroller's Office.* Report 1-2020, April.

Friedman, S., T. Insaf, T. Adeyeye, and J. W. Lee. 2021. Spatial Variation in COVID-19 Mortality in New York City and Its Relation to Communities of Color and Immigrants. Under Review. *Annals of Epidemiology*

Grady, S. C., and S. McLafferty. 2007. Segregation, Nativity, and Health: Reproductive Health Inequalities for Immigrant and Native-Born Black Women in New York City. *Urban Geography* 28 (4):377–397. doi:10.2747/0272-3638.28.4.377.

Hooper, M. W., A. M. Napoles, and E. J. Perez Stable. 2020. COVID-19 and Racial/Ethnic Disparities. *JAMA* 323 (24):2466–2467. doi:10.1001/jama.2020.8598.

Hotchkiss, M. 2015. Hypersegregated Cities Face Tough Road to Change. https://www.princeton.edu/news/2015/05/18/hypersegregated-cities-face-tough-road-change.

Kain, J. F. 2004. A Pioneer's Perspective on the Spatial Mismatch Literature. *Urban Studies* 41 (1):7–32. doi:10.1080/0042098032000155669.

Kawachi, I. 1999. Social Capital and Community Effects on Population and Individual Health. *Annals of the New York Academy of Sciences* 896 (1):120–130. doi:10.1111/j.1749-6632.1999.tb08110.x.

Kim, B., A. G. Rundle, A. T. S. Goodwin, C. N. Morrison, C. C. Branas, W. El-Sadr, and D. T. Duncan. 2021. COVID-19 Testing, Case, and Death Rates and Spatial Socio-Demographics in New York City: An Ecological Analysis as of June 2020. *Health & Place* 68:102539. doi:10.1016/j.healthplace.2021.102539.

Kramer, M. R., and C. R. Hogue. 2009. Is Segregation Bad for Your Health? *Epidemiologic Reviews* 31 (1):178–194. doi:10.1093/epirev/mxp001.

Krieger, N., J. T. Chen, P. D. Waterman, M. J. Soobader, S. V. Subramanian, and R. Carson. 2002. Geocoding and Monitoring of US Socioeconomic Inequalities in Mortality and Cancer Incidence: Does the Choice of Area-based Measure and Geographic Level Matter? *American Journal of Epidemiology* 156 (5):471–482. doi:10.1093/aje/kwf068.

LaVeist, T. A. 1993. Segregation, Poverty, and Empowerment: Health Consequences for African Americans. *Milbank Quarterly* 71 (1):41–64. doi:10.2307/3350274.

_____. 2003. Racial Segregation and Longevity among African Americans: An Individual-Level Analysis. *Health Services Research* 38 (6):1719–1734. doi:10.1111/j.1475-6773.2003.00199.x.

Leonowicz, A. 2006. Two-Variable Choropleth Maps as a Useful Tool for Visualization of Geographical Relationship. *Geografija* 42 (1):33–37.

Lieberman-Cribbin, W., S. Tuminello, R. M. Flores, and E. Taioli. 2020. Disparities in COVID-19 Testing and Positivity in New York City. *American Journal of Preventive Medicine* 59 (3):326–332. doi:10.1016/j.amepre.2020.06.005.

Lobo, A. P., and J. J. Salvo. 2013. *The Newest New Yorkers: Characteristics of the City's Foreign-born Population, 2013 Edition.* New York: New York City Department of City Planning.

Logan, J. R., and B. Stults. 2011. Diversity and Disparities: Residential Segregation by Race/Ethnicity – Sortable Lists of Segregation by Race/Ethnicity for the 200 Largest Cities. https://s4.ad.brown.edu/Projects/Diversity/SegCitySorting/Default.aspx.

Malega, R., and R. Y. Stallings. 2016. Regional Differences in Affluent Black and Affluent White Residential Outcomes. *Geographical Review* 106(1): 72–91. doi:10.1111/j.1931-0846.2015.12134.x.

Massey, D. S., and N. A. Denton. 1993. *American Apartheid: Segregation and the Making of the Underclass.* Cambridge, MA: Harvard University Press.

Massey, D. S., and J. Tannen. 2015. A Research Note on Trends in Black Hypersegregation. *Demography* 52:1025–1034. doi:10.1007/s13524-015-0381-6.

McLafferty, S. 2010. Placing Pandemics: Geographical Dimensions of Vulnerability and Spread. *Eurasian Geography and Economics* 51 (2):143–161.

McLafferty, S., M. Widener, R. Chakrabarti, and S. Grady. 2012. Ethnic Density and Maternal and Infant Health Inequalities: Bangladeshi Immigrant Women in New York City in the 1990s. *Annals of the Association of American Geographers* 102 (5):893–903. doi:10.1080/00045608.2012.674901.

NYC DOH [New York City Department of Health]. 2021. COVID-19: Data – Main Data Page. https://www1.nyc.gov/site/doh/covid/covid-19-data.page.

NYU Furman Center. 2018. *State of New York City's Housing and Neighborhoods in 2018.* New York: New York University.

Osypuk, T. L., A. V. D. Roux, C. Hadley, and N. R. Kandula. 2009. Are Immigrant Enclaves Healthy Places to Live? The Multi-Ethnic Study of Atherosclerosis. *Social Science & Medicine* 69 (1):110–120. doi:10.1016/j.socscimed.2009.04.010.

Phelan, J. C., and B. G. Link. 2015. Is Racism a Fundamental Cause of Inequalities in Health? *Annual Review of Sociology* 41:311–330. doi:10.1146/annurev-soc-073014-112305.

Pickett, K. E., and R. G. Wilkinson. 2008. People like Us: Ethnic Group Density Effects on Health. *Ethnicity & Health* 13 (4):321–334.

Quealy, K. 2020 The Richest Neighbohroods Emptied Out Most as Coronavirus Hit New York City. June 11 https://www.nytimes.com/interactive/2020/05/15/upshot/who-left-new-york-coronavirus.html

Rosenbaum, E., and S. Friedman. 2007. *The Housing Divide: How Generations of Immigrants Fare in New York's Housing Market.* New York: New York University Press.

Ssentongo, P., A. E. Ssentongo, E. S. Heilbrunn, D. M. Ba, and V. M. Chinchilli. 2020. Association of Cardiovascular Disease and 10 Other Pre-Existing Comorbidities with COVID-19 Mortality: A Systematic Review and Meta-Analysis. *Plos One* 15 (8):e0238215. doi:10.1371/journal.pone.0238215.

Stevens, J. 2015. *Bivariate Choropleth Maps: How-to Guide.* https://www.joshuastevens.net/cartography/make-a-bivariate-choropleth-map/.

U.S. Census Bureau. 2020a. Table B05002. Place of Birth by Nativity Status and Citizenship Status. 2019: American Community Survey 5-Year Estimates Detailed Tables – US and NYC. https://data.census.gov.

_____. 2020b. Tables B06004(B,D,H,I). Place of Birth in the United States. 2019: American Community Survey 5-Year Estimates Detailed Tables – US and NYC – For (Black or African American Alone; Asian Alone; White Alone, Not Hispanic or Latino); and Hispanic or Latino, Respectively. https://data.census.gov.

_____. 2020c. Tables B24010(B,D,H,I). Sex by Occupation for the Civilian Employed Population 16 Years and Over. 2019: American Community Survey 5-Year Estimates Detailed Tables –NYC – For (Black or African American Alone; Asian Alone; White Alone, Not Hispanic or Latino); and Hispanic or Latino, Respectively. https://data.census.gov.

_____. 2020d. Table B03002. Hispanic or Latino Origin by Race. 2019: American Community Survey 1-Year Estimates Detailed Tables. https://data.census.gov.

U.S. Centers for Disease Control and Prevention. 2020. *COVID-19 in Racial and Ethnic Minority Groups.* Atlanta, GA: Centers for Disease Control and Prevention.

USA Facts. 2021. US Coronavirus Cases and Deaths. https://usafacts.org/visualizations/coronavirus-covid-19-spread-map/.

Wadhera, R. K., P. Wadhera, P. Gaba, J. F. Figueroa, K. E. Joynt Maddox, R. W. Yeh, and C. Shen. 2020. Variation in COVID-19 Hospitalizations and Deaths across New York City Boroughs. *JAMA* 323 (21):2192–2195. doi:10.1001/jama.2020.7197.

White, K., and L. N. Borrell. 2011. Racial/Ethnic Residential Segregation: Framing the Context of Health Risk and Health Disparities. *Health & Place* 17 (2):438–448. doi:10.1016/j.healthplace.2010.12.002.

Williams, D. R., and C. Collins. 2001. Racial Residential Segregation: A Fundamental Cause of Racial Disparities in Health. *Public Health Reports* 116 (5):404–416. doi:10.1016/S0033-3549(04)50068-7.

Wong, D. W. S. 1997. Spatial Dependency of Segregation Indices. *The Canadian Geographer* 41 (2):128–136. doi:10.1111/j.1541-0064.1997.tb01153.x.

Wrigley-Field, E. 2020. US Racial Inequality May Be as Deadly as COVID-19. *Proceedings of the National Academy of Sciences* 117 (36):21854–21856. doi:10.1073/pnas.2014750117.

UNDERSTANDING THE SPATIAL PATCHWORK OF PREDICTIVE MODELING OF FIRST WAVE PANDEMIC DECISIONS BY U.S. GOVERNORS

PATRICIA SOLÍS, GAUTAM DASARATHY, PAVAN TURAGA, ALEXANDRIA DRAKE, KEVIN JATIN VORA, AKARSHAN SAJJA, ANKITH RAAMAN, SARBESWAR PRAHARAJ and ROBERT LATTUS

ABSTRACT. The uneven outcomes of the COVID-19 pandemic in the United States can be characterized by its patchwork patterns. Given a weak national coordinated response, state-level decisions offer an important frame for analysis. This article explores how such analysis invokes fundamental geographic challenges related to the modified areal unit problem, and results in scientific predictive models that behave differently in different states. We examined morbidity with respect to state-level policy decisions, by comparing the fit and significance of different types of predictive modeling using data from the first wave of 2020. Our research reflects upon public health literature, mathematical modeling, and geographic approaches in the wake of the underlying complex pattern of drivers, decisions, and their impact on public health outcomes state by statetime line. Contemplating these findings, we discuss the need to improve integration of fundamental geographic concepts to creatively develop modeling and interpretations across disciplines that offer value for both informing and holding accountable decision makers of the jurisdictions in which we live.

*W*hile virtually everyone in the United States has felt the direct or indirect effects of the COVID-19 pandemic, the pattern of impacts has unfolded unevenly across the country. Scientists of all disciplines have been working to make better sense of what has essentially come to be seen as an ever-changing patchwork of factors and outcomes. Inspired in part by a metaphor from a May 2020 article in *The Atlantic* entitled "We Live in a Patchwork Pandemic Now" (Yong 2020), our paper reflects upon this phenomenon and a growing public recognition in the early months of the pandemic that while national scale statistics first started to plateau, the underlying patterns of the spread were highly varied and dynamic at the state level, as well as at smaller county and city scales, going up in some places while diminishing in others. It strikes us that COVID-19 represents a compelling case to not only illustrate how modeling efforts, whether explanatory or predictive, suffer the unavoidable Modifiable Areal Unit Problem, or MAUP (Gehlke and Biehl 1934; Openshaw 1984), but also to uncover a unique arc of this story, where the real-time unfolding of decisions to respond to the pandemic reveal a spatial struggle (Leitner et al. 2008). Digging deeper, we can see how decision makers like governors, mayors, and state public health officials have issued a patchwork of policies, in absence of a cohesive national approach that forced the weight of the response to a particular intermediate scale, serendipitously opening up a space for our discourse about modeling and its real-time relationship to decisions to be examined and reflected upon. Similarly, public adoption and compliance of protective measures have unfolded unevenly over the nation, as information and trust on this issue reflect a generally fractured society. We furthermore add weight to this argument by assessing how patchwork characteristics can even be seen in the way that scientific predictive models behave in different locations, which is explored in this paper in analytical depth.

Researchers have turned attention and effort to applying their respective knowledge domains and collectively combining known techniques in hopes to innovate and respond with timely insights to inform public policy and influence public behavior during the pandemic. They have formulated and answered questions at many different spatial and temporal scales of analysis. Predictive models and mathematical forecasts have become the currency of academic monitoring and discussion, and ideally, informing the public and elected officials. Most studies are direct—they track infection rates, prevalence dynamics, morbidity or mortality patterns, and even excess deaths to follow the coronavirus across the country. Sometimes studies incorporate decision making into the models, such as the timing or appearance of policies to close activities like schools, restaurants and bars, gyms, or masking. More rarely, however, models may seek to account for both the spatiotemporal nature of decisions and the spatial-temporal behavior of people living in these jurisdictions where the decisions hold sway and are intended to influence their mobility and behavior.

We are interested in powerful machine-learning models, which were widely used as predictive tools for decision makers. In this article, we do not intend to test models. We do not intend to provide an answer about which predictive approach is "best." Instead, we hope to structure a broader, interdisciplinary discussion unraveling the spatial incongruence among these contributing elements. We seek to explore how the human-physical complex system of a pandemic—including behavioral, decision making, and epidemiological data—may be inconsistent and incongruent with each other, and thus point to the centrality of fundamental geographic challenges in the production of knowledge about the pandemic. The importance of this framework is to underscore the need for explicit, well-justified attention to choices of spatio-temporal scale, relevant to the (real-time) decision-making context, to potentially improve transferability of findings (but not necessarily models) from one place to another, as well as to offer a measure of accountability for evidence-based decisions at the scales in which they are made. Ultimately, these insights hold true not only for outbreaks such as COVID-19, but also for understanding and responding to other public health concerns, such as disaster response or future infectious outbreaks.

To reiterate, the purpose of this exploratory discussion paper is not so much to identify a good model or assess a good modeling method, but the main purpose is to reflect upon fundamental geographic principles in the process of the scientific modeling in a real-time complex global pandemic in ways that inform decisionmakers and the public. We raise interdisciplinary awareness of such analytical challenges, and conclude with some recommendations on what we as a scientific community may learn from the first wave experience of COVID-19.

BACKGROUND AND LITERATURE

Actions in the early days and weeks of the COVID-19 pandemic taken to mitigate spread disproportionately affects long-run impacts on public health. Such actions and consequences always entail spatiotemporal components, but linking certain actions to consequences is generally not straightforward in complex human-physical systems, such as a pandemic. The beginning of the COVID-19 pandemic in the United States witnessed a deafening lack of a unified federal response, including mismanagement in controlling the border, lack of workplace standards (Hanage et al. 2020), and not quickly ensuring enough primary protective equipment (PPE) for essential workplaces (Lagu et al. 2020). Additionally, the limited response that did come out of the federal government was slow, and at times confusing (Haffajee and Mello 2020). To make matters worse, the federal government failed to look at evidence-based practices learned from previous outbreaks and disasters to inform the guidelines that did make it to the public (Solinas-Saunders 2020). These sentiments led most Americans to think the United States' national response did a poor job at addressing the pandemic (Pew Research Center 2020). It also forced an interesting natural experiment on the scientific modeling community—to reckon with a set of actors—governors—to monitor and predict a novel coronavirus and the impacts of decisions that were made at this scale in absence of federal powers that typically provide some measure of coordination across states.

To provide context for this study, we first lay out the backdrop of public health dynamics, both in terms of COVID-19 and pandemics more broadly from the public health literature. The varied nature of virus transmission and resulting patterns of cases followed by deaths are in part attributable to the highly contagious character of the novel coronavirus, as well as physiological responses of infected individuals.

Secondly, we lay out a few relevant principles from the perspective of predictive mathematical modeling domains to provide some structure to assess the many attempts to understand moving parts of the pandemic. This also introduces the choice of methodologies explained later that illustrate the challenge of the MAUP.

Finally, we consider fundamental geographic problems within decision making contexts. Clearly the behavior of people—how they move and what precautions they take—matters a great deal, impacting upon the resulting spatial distribution of the disease. These individual decisions are affected both by advice taken directly from public health officials and by the policies put in place by decision makers with jurisdiction to restrict or require mediating activities over the landscape of places where people work and live.

PUBLIC HEALTH CONTEXT

After COVID-19 was declared a national public health emergency by the health and human services secretary on 31 January 2020 (U.S. Department of Health and Human Services 2020), U.S. states began taking their own approaches to address the pandemic within their borders. The response to such widespread outbreaks is by law the joint responsibility of state governors and the federal government. Some states took aggressive immediate measures to limit the spread of COVID-19, for example, declaring a statewide public health emergency or developing a coronavirus task force. Since the turn of the century, the United States has responded to a number of pandemics including Ebola (Gostin et al. 2014; SteelFisher et al. 2015), swine flu (Butler 2009; Coker 2009), and SARS (Park and others 2004; Rothstein 2015). In these responses, researchers were able to assess policy responses to limit disease spread in near real time. Preparation plans are considered a major factor in influenza pandemic preparation, however, these plans must be constantly adjusted and updated based on lessons learned (Leung and Nicoll 2010). Others argue that proper plan execution is more important than the plan itself (Gibbs and Soares 2005). A stark example is found in critiques of responses to Hurricane Katrina. The hurricane itself exposed breakdowns in the chain of support for disaster relief, leaving communities that were already vulnerable in a state of heightened vulnerability (Quinn 2006). Many scholars noted extreme consequences of failing to improve emergency response efforts to disasters after Hurricane Katrina (Schneider 2005; Sobel and Leeson 2006; Holguín-Veras et al. 2007) saying, "unless we take to heart the lessons that Katrina teaches, especially improved systems for communication and coordination, we are likely to repeat the Katrina problems" (Kettl 2006). The same day of the first Ebola-related death in the United States in 2014, the Center for Disease Control announced increased screenings at JFK, Washington-Dulles, Newark, Atlanta, and O'Hare airports (Gostin et al. 2014) in response to the fact that 94 percent of people coming from Sierra Leone, Guinea, and Liberia (Ebola-affected nations) travel through these five airports (Center for Disease Control [CDC] 2014). This exemplifies a federal-level policy that affected state-level prevention efforts taking spatiotemporal realities into explicit account. Similarly, the federal government began an epidemiological surveillance system during the SARS pandemic that stemmed from state and local health departments working to report cases to the CDC (Schrag and others 2004). After declaration of the public health emergency in response to COVID-19, however, the robustness of a federal response seen in previous epidemics was absent. Except for some travel restrictions, minimal federal policies regulated activity to slow the number of new COVID-19 infections within national borders (Haffajee and Mello 2020; Gostin et al. 2020).

From the official White House "30 Days to Slow the Spread" report, suggestions on limiting COVID-19 cases were provided, but initial instructions tell residents to "listen to and follow the directions of your state and local authorities"; (The White House 2020). Delegating the bulk of the responsibility to the states to make COVID-19-related decisions precipitated a wide spectrum of state and local level policies, that we suspect further contributed to the patchwork character of the pandemic.

While the literature addresses the spatial variability of factors that lead to disease spread, we find limited scholarly work focused on patchwork policies in public health. Research on the opioid epidemic shows that while the epidemic has been a national emergency since 2017, the epidemic looks different depending on geographic location; (Rigg and Monnat 2015), socioeconomic status (Altekruse et al. 2020), race/ethnicity (Pletcher and others 2008; Alexander et al. 2018), age (Campbell et al. 2010), and gender (Choo et al. 2014; Graziani and Nisticò 2016). Some studies show that prescription-opioid misuse is more common in rural areas compared to urban centers (Keyes et al. 2014). Some studies reveal how federal policies on opioids need to not only be flexible, but also account for variations in state or local realities (Chakravarthy et al. 2011; Nelson et al. 2015). While sensitive to the geography of factors and of outcomes, these studies fall short of analyzing how the spatial variability of policies and choice of spatial unit of analysis relate to accountable decision making.

Journalistic coverage of COVID-19 highlights the need for understanding the ways outbreaks operate as a patchwork over space. *The New York Times* used data from the University of Oxford to visualize varied outcomes, concluding, "the surge is worst now in places where leaders neglected to keep up forceful virus containment efforts or failed to implement basic measures like mask mandates in the first place" (Leatherby and Harris 2020).

Swift responses are essential (Mallinson 2020; Pikoulis and others 2020), but the ability to project future outcomes is also a key component of preparedness. Examples of this include the response to the September 11, 2001, terrorist attacks future implications on mental health and environmental effects (Rosenfield et al. 2002; Reibman et al. 2016); the 2003 outbreak of SARS (Smith 2006; Krumkamp et al. 2009), the 2014–2016 Ebola outbreak (Maffioli 2020), or H1N1 in 2009

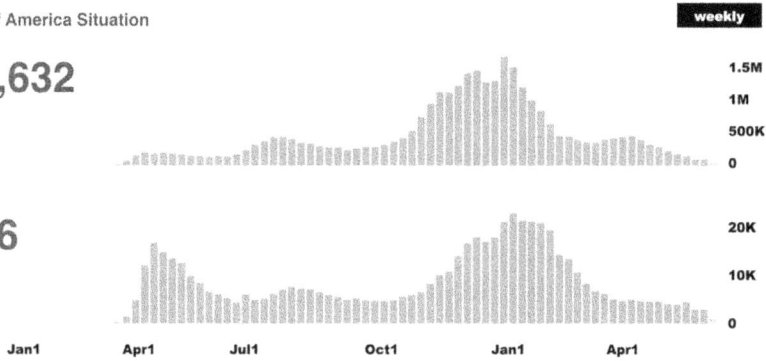

FIG. 1—COVID-19 Cases and Deaths for the entire United States, January 2020 to June 2021; Source: World Health Organization, Dashboard at https://covid19.who.int/region/amro/country/us

(Fineberg 2014), as well as efforts to minimize the impact of annual illnesses like the flu (Thomson et al. 2018). The scientific scramble to respond quickly and with predictive power to the COVID-19 pandemic is exemplified in the sheer number of rapid research grants awarded. The National Institutes of Health issued over 100 grants (National Institutes of Health 2019) and the National Science Foundation awarded roughly 900 grants (National Science Foundation 2019) to address various components of the COVID-19 pandemic.

As the United States might nationally be seen as riding out the "third wave," (see Figure 1), we emphasize that outcomes rely both on the public health decisions and mandates put in place to protect people, as well as their compliance. These vary by place. This national pattern continues to belie state-to-state variance in outcomes and the unfolding of many pandemics. This pattern adeptly illustrates the MAUP. In the following section, we consider how modeling efforts are confounded by this persistent analytical challenge.

MODELING CONTEXT

There has been a tremendous amount of interest in mathematical modeling and forecasting of the COVID-19 pandemic outcomes. Modeling approaches can be grouped into one of two general categories: forecasting models and mechanistic models. Forecasting models are typically statistical in nature and attempt to predict patterns in the data; where predictions "de-noise" the data and learn some latent representation of the phenomenon over time. Such models have been used both in the context of modeling past infectious diseases and in the context of predicting features of the ongoing COVID-19 outbreak. For instance, (Kane et al. 2014) use time-series modeling techniques (Box et al. 2011) to study past (influenza-like) outbreaks. Ceylan (2020) are examples of works that use time-series modeling techniques specifically in the case of COVID-19. Several complementary machine-learning-based sequence models have also been proposed in these scenarios (C. Wang and others 2020; Yang et al. 2020; Wang et al. 2020c).

On the other hand, mechanistic models (such as the SEIR model) attempt to directly understand and exploit the mechanism of the underlying virus under various disease-specific assumptions. There has been a flurry of activity in adapting such models (Wu et al. 2020). Others are hybrids between mechanistic and forecasting models (Reiner et al. 2020). Probability-based COVID future risk estimation approaches like COSRE (Sun and others 2020) provide risk estimation on a county level with strong reliability, supporting day-to-day decisions like risks related to spatial behavior and mobility to inform individual activities.

Regardless of choice among these models, they each require definition of time scale, spatial scale, and a scope of data to frame their application, and are susceptible to the MAUP. This holds implications for understanding past and future accountable decision-making.

DECISION-MAKING CONTEXT

Researchers, applying various models to understand this complex unfolding of the pandemic, will discover patchwork characteristics across spatial units of analysis. By this, we mean that while one would normally expect adherence to such standard geographical phenomena as Tobler's First Law of Geography (1970), patterns of the COVID-19 pandemic do not conform neatly, and how to frame which models is not an obvious exercise—especially in an interdisciplinary setting. In other words, near things might not in fact be more related than distant things when the default jurisdiction of public health decisions is relegated to each governor of every state. As we argue in this paper, given the way that public health outcomes rely on complex systems of spatiotemporal behavior—comprised both of individuals and of the special decision-making behavior of officials that influence individuals there is underlying patchworkiness, or nonconformity, that can be observed beyond the diversity in state summary statistics of raw health outcomes like prevalence or morbidity. As we will discuss, the modeling performance itself exhibits patchwork results.

The variability and diversity of patterns of the pandemic(s) are revealed when researchers try to create data-driven predictive models of disease statistics in the presence of behavioral measurements such as mobility, mask-wearing, and activity-opening decisions. Our work specifically suggests that these observations beyond the obvious patchworkiness in measurable public health statistics are also reflected significantly in the patchwork patterns of decisions that in turn rely on the monitoring output and performance of predictive models created to explain the pandemic. We interpret this as evidence of the critical importance of the scale incongruence of decision making in responding to spatial phenomenon like the fast-moving coronavirus pandemic.

This observation builds upon a unique geographical problem first introduced in the *Geographical Review*, that recognizes the special nature of types of spatial decision-making behavior that both incorporate and influence individual decision-making behaviors (Solís et al. 2017). This Decision Accountability Spatial Incongruence Program (DASIP) is a special cause of the MAUP, which draws particular attention to the agency of special actors and sets of actors (elected officials, public administrators, stakeholders, CEOs) who are responsible for jurisdictions that potentially shape or, in the case of nonconforming behaviors such as protesting pandemic remedies, informs outcomes that in turn can affect others in important ways. This paper seeks to underscore the importance of the spatial scale of such decisions (for example, closures and mask mandates), and helps to shed light on how they may be incongruent with the desired outcomes (cases, deaths) in order to open interdisciplinary scholarly discussion and reflection on the unfolding of COVID-19. Furthermore, this framework helps us to ask whether and how decision makers can be held accountable in ways that are more congruent with the data, decisions, or impact.

As noted above, a gap in the public health and epidemic modeling literature so far is thus the explicit inclusion of study elements that focus on the role and scale of decision making as it relates both to this behavior and the outcomes of the pandemic. Furthermore, research that illustrates such problems by interrogating the spatial performance of the models incorporating decisions on COVID-19 is rare, if not nonexistent.

We seek to fill in some parts of this important gap by exploring not only the patchwork character of how decision making about the pandemic has unfolded, but also the patchwork character of model performance to make clear this geographical problem. This reflects upon how the research community has come to understand it, choosing to illustrate this at the state-level as one important key unit of analysis in this case. This is justified given that public health decisions were largely delegated from the federal level to governors (Djulbegovic et al. 2020; Jacobson et al. 2020).

The time period of analysis runs from January through September 2020, largely covering what is seen nationally as the first wave, representing a critical moment in the subsequent ability to manage the pandemic in later stages. By explicitly revealing the patchwork character of some categories of predictive modeling, we open a reflection on our role as a scientific community seeking to support monitoring and response in a context of this pandemic.

METHODOLOGICAL FRAMEWORK

We illustrate how this fundamental spatial challenge played out in the COVID-19 first wave by devising a multifactor, machine-learning model to fit one state, Arizona.[1] We then test the extent of this state-model fit to the other 49 states in terms of how well or poorly it can describe the spatiotemporal pattern found in other states.

To test this idea, we choose several simple representative modeling techniques to perform our analyses. While there has been some recent work; (Kang et al. 2020; Mollalo et al. 2020; Li et al. 2020) in understanding the spatial statistical behavior of COVID-19, these studies typically look at the spatial distribution of raw data/fundamental public health outcomes/corresponding to various metrics of interest related to COVID-19. To the best of our knowledge, ours is the first study that endeavors to do the same using the behavior of models themselves.

The models were built from an increasingly complex scaffolding of data layers in order to illustrate the persistent impact of the MAUP. We began modeling raw outcomes of the pandemic (without decisions[2]), that took changes in daily deaths to predict changes in deaths at a future date. We then added data to reflect decisions about closures, using mobility data as an indicator of the spatial behavior of residents, together with the morbidity data. This was used to understand at what future date changes in public health outcomes would occur, and computed changes in those outcomes themselves. Finally, we added consideration of nonmobile decisions regarding face-mask mandates, as a second policy constraint. For each state, a measure of fit was noted. We operationalize the idea of what we call "patchworkiness" by calculating the spatial autocorrelation of this measure of fit by state. The degree of spatial clustering and its related significance represents what character of patchwork the model can reveal.

DESCRIPTION OF THE DATA USED

The first step to developing a patchwork model relied on depicting a variety and time line of state-level policies. To do this, a time line by state spreadsheet served as the foundation to record and assess the patterns of state level policy decisions. The columns represented dates while rows represented all 50 states, plus the District of Columbia and Puerto Rico. The 52 geographic areas were coded into 6 codes: mask mandate, stay-at-home order, social distance/gathering limitations, antirestriction policies, other, and end of restrictions. "Other" served as a catchall code that included a state's first confirmed case, a declaration of a state public health emergency, when testing began, and several other policy decisions. When a state had a COVID-19 related policy go into effect, that policy would then be noted in the corresponding date column to the day that policy went into effect.

The "mask mandate" code was used whenever a statewide mask mandate went into effect. Not all states implemented mask mandates, and this code was only used for statewide policies and not policies enacted at the county or municipality level. The "stay at home order" code was used whenever governors enacted restrictions of curfews on people's movements within the state. This was also sometimes referred to as shelter-in-place, depending on the state. The "social distance/gathering limitations" code included statewide restrictions on the number of people that could gather at once as well as requirements on keeping a certain amount of distance between people. This also included school closures, cancelation of elective surgeries, and limiting dine-in options at restaurants. "Antirestriction policy" code referred to the instances where governors or state officials prohibited lower-level decision making within the state. For example, on 27 April 2020, Governor Greg Abbot of Texas issued an executive order making local officials unable to enforce a mask mandate or impose any mask related fines. The "other" served as a catchall of pertinent COVID-19 related information, rather than policies. This included relevant information such as the date of the first confirmed COVID-19 case in the state and the date of a state-level declaration of a public health emergency. Finally, the last category, "end of restrictions," chronicled the time line of states reopening. This included the reopening of gyms, restaurants, and other business and the lifting of stay-at-home orders. This spreadsheet[2] consolidated general time lines at state-level policies ranging from 21 January 2020 to 15 August 2020, representing first wave time lines.

To explore the patchwork of models, we choose to use normalized reported deaths as the base data on pandemic outcomes to work with. While there are various options for raw sources of data, we used data available from CDC for morbidity ("CDC COVID Data Tracker" 2020); from the Descartes Lab for mobility (Warren and Skillman 2020); and Blavatnik School of Government for government response, (Hale et al. 2021) in addition to triangulating with our own decisions-by-state tracker described above. The data for our analyses are obtained by first merging the records from CDC and Descartes Lab such that we obtain aligned time series for deaths as well as mobility. For a given day, available information consists of deaths and associated mobility indices for that day for each state in the United States (omitting Puerto Rico and the District of Columbia in the models).

We experimented with different model classes, namely, decision trees, time-series models, and the like, where the input is either a snapshot or contiguous segment of changes in cases, mobility, and deaths, and the output is predicted changes in death at future time. The modeling methodologies of our paper are based on developing statistical forecasters of state-level mortality using past mortality and mobility data.

With our experiments we do not showcase ways of improving existing methods. Instead, we point out how modeling developed at a particular spatial unit performs at other spatial units, to reveal the dependency on geographic principles. The experiments have validated the idea that the variables (model parameters) governing the spread of COVID-19 are different for each state (they depend on the choice of scale and extent).

Our next methodological aim was to quantify and evaluate measures of model variability across geospatial locations. As a first step toward this, we developed a well-tuned forecasting model for one state: Arizona, which in addition to being a convenience choice, has the distinction of being among the top three states of earliest confirmed COVID-19 case, with the longest time lag to a first decision to mediate. We then recorded the performance of this model for other states using an R^2 prediction score. That is, for each state a, we compute a score $s(a)$ that captures the performance of Arizona's forecasting model for state a's data, which is simply the R^2 computed for state a using Arizona's model. We then estimate the spatial variability of $s(a)$ across the states using the standard measure of spatial autocorrelation, Moran's I (Moran 1947, 1948). Moran's I for the scores $s(a)$ is defined as follows:

$$I = \frac{N}{W} \sum_a \sum_{a'} w_{aa'}(s(a) - \bar{s})(s(a') - \bar{s}) / \sum_a (s(a) - \bar{s})^2,$$

where N is the total number of states being considered (50 here, where we omit Puerto Rico and the District of Columbia due to partial data and special status. Thus, $\bar{s}$ is the average score across the states, $w_{aa'}$ is the spatial weight for the pair of states a and a' (using a Euclidean distance-based measure). Comparing the value of I with its expected value (which is $-1/N$ - 1 = - 0.021 here) allows us to estimate the amount of spatial variability of the scores. These measures operationalize and demonstrate our experimental concept of a patchwork pandemic.

ANALYSIS

To deal with possible missing data in the time series, we performed basic linear interpolation for removing gaps in data. We then perform a linear de-trending of the data by working with the first derivative of all the available time series. For a given day, which we will denote as t, we will let d_t denote the number of deaths on day t. We will also let $m_t^{(1)}$ and $m_t^{(2)}$ denote the value of the mobility indices from the Descartes Labs dataset. We further process the data so that the model input variables correspond to changes in cases, deaths, and changes in the various mobility indices. For a given day, denoted by t, the corresponding features are $X_t = [c_t, d_t, m_t^{(1)}, m_t^{(2)}]$. Then, the model output is given by $Y_t = [d_{t+\Delta}]$, where $\Delta > 0$ is the prediction horizon, that is how far in the future our model is asked to predict. In addition to this, we also analyze the results by adding a feature f_t that denotes facial-covering index. Facial covering index is an ordinal value on the 0–4 scale, as defined by (Hale et al. 2021) to represent how strict the facial covering policy is. As a result input tuple $X_t = [c_t, d_t, m_t^{(1)}, m_t^{(2)}, f_t]$ and the model output is given by $Y_t = [d_{t+\Delta}]$.

We deployed four distinct analytical pieces of evidence to characterize the patchwork patterns of the pandemic, relative to the baseline state:

RANDOM FOREST MODEL

In the first model, we used a random forest regressor (Liaw and others 2002) that estimates a function $Y_t = f(X_t)$. A decision-forest based forecaster was fit on Arizona data corresponding to predicting future mortality from data of past mortality and mobility. Decision-forests are considered a standard approach in predictive modeling, thus we started with this choice of modeling method. This devised Arizona model was then tested on other states. To train the model for a given state, we randomly sample 70 percent of the data for training, and reserve 30 percent for testing. We used the implementation from the package sklearn (Pedregosa et al. 2011) with the default number of trees = 100 and max depth of each tree = 2. We first fit an optimal model to the time-series data measured from Arizona. We then used this optimal model for Arizona to attempt a forecast of mortality across all the other 49 states and record the corresponding R^2 score, which is a quantification of the predictive quality of the model. This geospatial spread of R^2 scores is then analyzed using a standard spatial autocorrelation test (Global Moran's I) to compute how spatially variable the scores are.

Auto-regressive model: In the second variation, we developed an analogous approach as above, but this time, we used time series based forecaster that models a window of observations in the past as the basis for predicting a single outcome variable in the future. The technique we used was an auto-regressive moving average (ARMA) model, which fits a linear function $Y_t = f(X_{t:t-\delta}, Y_{t-1:\tau-})$, where $X_{t:t-\delta}$ is the time-series data corresponding to the variable X between times $t - \delta$ and t; we chose the hyperparameters to be $\delta = 5, = 1$. These choices can be made using an information criterion such as AIC (Bozdogan 1987) as more data becomes accessible from the pandemic. We again choose the best parameters for fitting an ARMA model at Arizona, forecast mortality in other states, record the R^2 score for each state and compute the corresponding Global Moran's I score.

GRANGER CAUSATION MODEL

The models as described above use one state as a base model, Arizona. Data from other states were used to test the model. Ideally to test the patchwork concept, we would like to compare models by estimating them individually per state. This is hard to do since it is nontrivial to compare model-to-model mismatch as compared to model-to-data fit (spatial comparison adds further complexity). In order to bridge the gap between our approach above and this ideal, for each state, we consider another notion of forecasting performance—Granger causality (Ding et al. 2006). In this analysis, we estimate the Granger causation between the mobility index time series $m_t^{(1)}$ $\left(or m_t^{(2)}\right)$ and the mortality time series d_t. Granger causality is computed by measuring the improvement in the predictability of the mortality time series when one includes the mobility time series as part of the time-series fit. For each state, we now optimally tune its own hyperparameter (lag for Granger causality) and choose the respective chi-squared test statistic of highest significance (chosen from the values computed for a series of lags). This value is used again to compute Moran's I of the joint mobility-mortality models.

POLICY MODEL

To depict multiple decision points in one of the models, we incorporated a facial-covering index as a feature in the dataset, permitting us to analyze the impact of policies on our model. We repeated a similar experiment as in the first case where the Arizona model is fit to all other states to obtain an R^2 score. All other aspects of the modeling remain the same as the first random forest model described above barring a change in the input data. The input data on facial coverings is a time series of values on an ordinal scale (0–4), with 4 being the highest degree of enforcement for each state.

SUMMARY OF RESULTS

States, given the imperative to implement pandemic policy in the context of a lack of national coordination, began slowly after the first appearances of confirmed COVID-19 cases, to respond, typically by declaring a public health emergency (Figure 2). Only a few states declared a public health emergency prior to their first known case. Typically, weeks later, the first stay-at-home orders began to emerge, generally followed by mask mandates, when they existed. Figure 2 provides an overview of this sequence for all 50 states (plus Puerto Rico and the District of Columbia), although the conflicts between federal and state policy, and state and county or city policy is not depicted. Despite that governors more or less followed a typical order of action, the timing of these decisions varied during the first wave of the pandemic, and showed little relation to political party of the governor.

In terms of modeling public health outcomes relative to behavior and decisions, the results from different experiments are summarized in Table 1. The "Model" column represents the algorithm used; the "Input" and "Output" columns summarize what a particular model outputs were given based on the

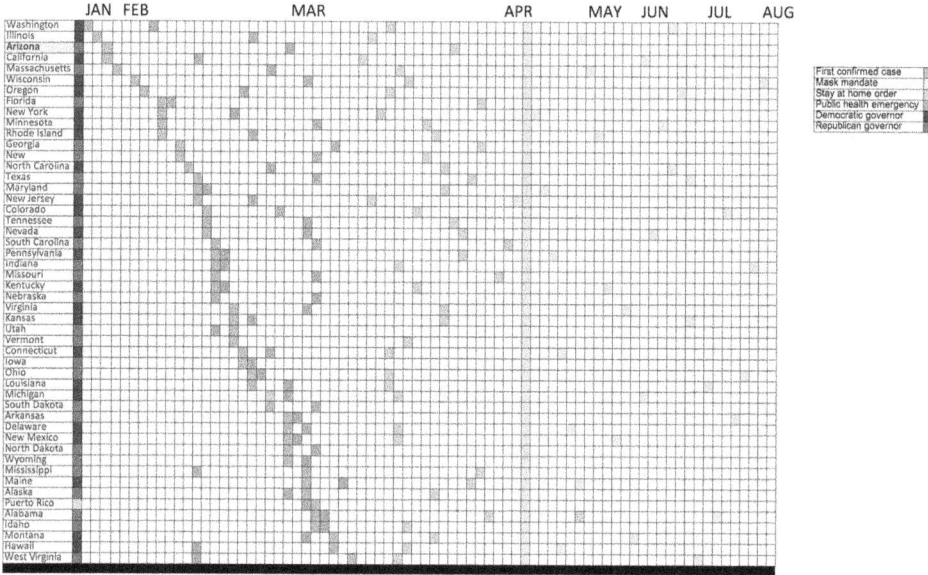

Fig. 2—Visualized time line of decisions by state, January—August 2020.

inputs. For each model we calculated a value representative of each state's quality of fit (to the Arizona model) using the metric specified in the "Measure of model fit" column. These representative values are used as an input to the calculation of spatial autocorrelation whose results are summarized under p-value and Moran's I. We found similar results using population normalized deaths. Based on these results, we find that patchworkiness using the random forest regressor with depth 2 was consistent across methods.

When mapped by quality of fit by state, and spatial clustering run, patchwork patterns remained regardless of whether the random forest model, ARMA model, or the Granger model was used to visualize clusters (Figure 3).

This map might be interpreted to depict places with similar differences to the base Arizona model, relatively speaking, and may indicate where the pattern of cases and deaths were similar in similar time ranges, where people behave similarly, or where governors may have even worked together on joint policy. As an experiment in visualizing patchwork patterns resulting from behavior and decisions, the map provides inspiration to consider different levels of conformity and/or fragmentation, despite different significance of spatial mismatch.

DISCUSSION

In this study we explored spatial and temporal incongruence among the way that the natural phenomenon of the health pandemic unfolded, the behavior of people via mobility, and the decision makers choices responsible for regulating

TABLE 1—A SUMMARY OF MODELS TESTED, WITH CORRESPONDING INPUTS, OUTPUTS, MEASURES OF FIT, OBTAINED MORAN'S INDEX AS AN INDICATOR OF PATCHWORKINESS, AND THE SIGNIFICANCE LEVEL

MODEL	INPUT	OUTPUT	MEASURE MAPPED	MORAN'S I (EXPECTED VALUE: −0.021)	P-VALUE (SIGNIFICANCE LEVEL > 0.05)*
Random Forest	Changes in daily mobility, changes in cases, changes in daily deaths	Changes in deaths at a future date	R^2 value	0.0332	0.0392*
Random Forest	Daily mobility, Population normalized changes in cases, Population normalized changes in daily deaths	Population normalized changes in deaths at a future date	R^2 value	−0.154	0.1513
ARMA	Changes in daily deaths	Changes in deaths at a future date	R^2 value	0.1726	0.01737*
Granger Causality	Changes in daily mobility, changes in daily deaths	Chi-squared statistic for a series of lags	Chi-squared statistic of lag with highest significance	−0.1288	0.2636
Policy Model	Ordinal scale of facial coverings; Changes in daily mobility, changes in daily deaths	Changes in deaths at a future date	R^2 value	0.01	0.51

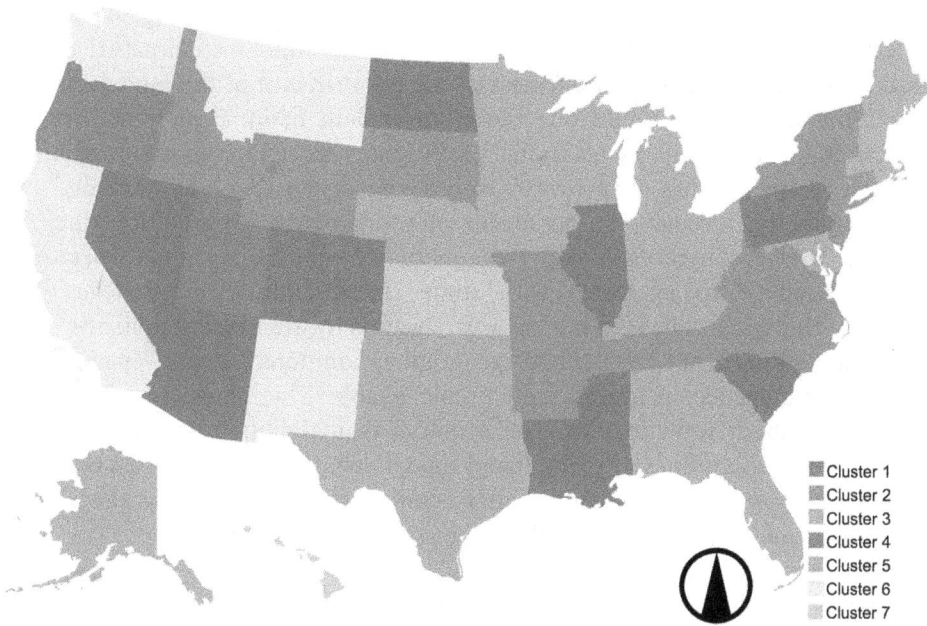

Fig. 3—Cluster Map results of state-wise modeling depicting patchwork relative to the Arizona base model.

and guiding resilience responses to these events, in the serendipitous context of lack of overall national coordinated response. Certainly, the whole story of the pandemic in any place, at whatever scale or spatial unit, cannot be fully told through modeling, and as a representation of the phenomenon, neither can public-facing or even published research fully convey complexities. These partial findings hint at how the resulting patchwork runs deeper than the respective patterns of public health outcomes, but also for our models and understanding of it. We find that designing a base model and comparing it to other states reveals dimensions of match and mismatch. And while the fit varies, significance of all models do not hold universally, modeling persistently shows a patchwork of clusters, beyond just state-by-state decision making. The fact that the patchwork did not conform only to state lines confirms that factors other than governors' decisions still retain an importance for understanding how the pandemic unfolded.

Assessing the model with respect to how well the decision making relates back to behavioral-dependent outcomes affords a framework to ask questions about accountability. The very least of inquiry seeks to understand the spatial and temporal mismatch among the decisions made, outcomes, and accountability for those outcomes, ideas which we began to explore with this work.

We tested the model with respect to how well it does or does not describe these relationships at the state level, not by mapping the patchiness of outputs (cases, deaths) or drivers (mobility), but by graphing the fit.[3] Doing so helps us reflect on our role as scholars in making sense of the fundamental nature of this kind of pandemic: the patchworkiness of the performance of the models underscores the need to not build one model to explain national patterns, but the critical importance of many researchers from many disciplines in many locations building many models that reveal aspects of society, decision making, and behavior in the face of risk, cognizant of fundamental geographic principles like the MAUP. These explanatory or predictive models can often operate in field focused teams or teams with some interdisciplinary approaches. For instance, the team collaborating on this study is vastly interdisciplinary, with expertise coming from the fields of geography, computer science, engineering, artificial intelligence, planning, and public health. This allowed for complex discussions about ways to formulate a model that highlights research practice from multiple disciplines. It precipitated robust debates about the most useful or appropriate spatial and temporal scale to contribute to adding new knowledge, even about the meanings of usefulness or appropriateness.

Limitations of using this set of machine learning modeling approaches to try to uncover relationships reveal potential areas for future research needs. For one, the stand-in data for behavior (and spatial behavior) in terms of mobility is incomplete. The data would not represent the full mobility of the population, nor would it fully be expected to estimate compliance with stay-at-home orders. Compliance with mask mandates is even more uncertain in these data, and such slippage in behavior certainly accounts for some of the spatial incongruence of the models, as well as for the ultimate health outcomes. In other words, some people in some states (or regions) will just comply better.

Furthermore, even the best metrics of social behavior cannot fully account for persistent social vulnerabilities (Wang et al. 2020), which may structure behavior, but we do not portend is conflated with it. Future studies could enable factoring in some of the known vulnerabilities by demographic (age, race, income) or some other empirically derived weighted social vulnerability index, perhaps even at a smaller spatial scale, as the state unit used in this analysis is likely too coarse to overcome this limitation. We believe our illustration of the underlying spatial analytical problem would still be present, no matter how many factors such modeling contains, or how well one model fits to a particular choice of spatial unit (Fotheringham et al. 2017; Li and Fotheringham 2020).

Similarly, findings from more focused studies, or comparative studies like Praharaj et al. (2020) could improve modeling of mobility data to measure the effect of COVID-19 policies on the specific changes observed. While our experiment accommodated a range of data, the models barely accommodated the burgeoning body of knowledge around COVID-19, something that future elaboration of these ideas should do.

The results from this research shows that the Arizona test model fits better in some places and worse in others. Regionally, one might expect the situation with COVID-19 to follow similar patterns (for example, California and Arizona would have more similarities compared to Arizona and Massachusetts). Even when factoring for spatial autocorrelation, the differences in the model still seem to be patchworky. In terms of applicability to public health, this further confirms what public health scholars have already noted: illnesses and diseases can vary based on geographic location, policies, and behavior. We can also see this notion in the example of opioid epidemic noted previously. Moving away from a one size fits all approach to predictive modeling could allow for the capture of these important health behavior, exposure, and healthcare access nuances that drive disease and illness variation.

Finally, the challenge looms about where decisions are made and whether decision makers can accommodate such complexity in their actions. While it was quickly clear that there would be scarce national scale response, and that governors must assume larger than usual responsibility for policy, future research should explore mechanisms to test for the contradictions across variable scaled decisions, decisions that are incongruent with their jurisdictions, and simultaneous decision-making conflicts, since examples of governors who struggled with mayors, and other policy actors abound (Leitner et al. 2008).

It bears repeating that is not our intention in this paper to create a new model or a new class of modeling. We instead seek to illustrate how the patchworkiness in the spatial performance of these different models that seek to explain the pandemic, and in turn serve as real-time, decision-making tools, itself should be a source of reflection in the overall assessment of how it all unfolded and scientists' role in that process.

In the end, this work showed patchworkiness at a deeper level than that which may be apparent from summary statistics of cases, hospitalizations, or mortality. Instead, we found patchworky behavior in the way predictive models behave in different locations. One way to state this is that we found that when we develop a predictive model, based on past mortality and mobility and policy data, to predict changes in future pandemic related mortality, there is significant model-mismatch of the performance of such a model across geographic locations and at different spatial units. We saw that this finding carries over across a few different variations of whichever underlying predictive modeling paradigm is used, specifically decision forests and auto-regressive models.

Based on these results, we suggest the following recommendations for future research: engaging in interdisciplinary teams, recognition of scale quickly, closer ties to actual decision makers, and reinforcing the contextual nature of geography even in heavy data modeling research. In the context of the COVID-19 pandemic, more geographers working with public health professionals, data scientists, and policy scholars could help amplify and broaden awareness of the fundamental geographic challenges as events such as this pandemic happens in real time. One of the downfalls of the response to the COVID-19 pandemic was the lack of widespread recognition that the performance of predictive models would manifest differently among and within states. Having a quicker recognition of the importance of understanding scale choice in monitoring and predicting behavior of future outbreaks and natural disasters can improve scientists' response to make predictive models that overcome the MAUP. The patchworkiness captured in this study also presents the need for more joint reflection on these challenges between decision makers and data scientists.

CONCLUSIONS

A more coordinated, national-level response may mark future interventions for both health-related outbreaks and natural disasters (hurricanes, tornados, and the like), given the prospect of a new federal administration, but the ineffectiveness of a patchwork response in the early weeks and months cannot be underestimated. This exploration reiterates the need to innovate methodologically around decision making as a particular category of spatial behavior, and pandemic patchwork policy making as a specific instance of spatiotemporal, scaled decision making that unfolds differently across jurisdictions, and according to different modeling performances. Our results imply the spatiality of collective, accountable decision-making behaviors as distinct from independent, individual choices. But it also reinforces the connections among policy and compliance/noncompliance, something that this particular pandemic and the complexity of mediating actions brought into stark focus. Future research is needed that involves interdisciplinary teams of scientists engaged reflexively in practice together with decision makers to explore methodological solutions to the modifiable areal unit problem, and to the decision-making accountability spatial incongruence problem. The outcome of geographic contextualization as well as deliberate and justified choice of spatial scale and units would ideally be better support to real-time, evidence-based modeling of the impact of decisions and public accountability for those decisions—as such emergencies as the COVID-19 unfolds.

We conclude with the need for explicit attention toward the integration of researchers who use predictive modeling to proceed at various scales, and work with the multiple scales of decisions and decision makers. While this conclusion on first glance may seem trivial, or superfluous, in reality, the practices of both decision makers and scientists seeking to understand phenomena, especially in a rapid response context, does not always follow this "first mile" best practice. Some of this modeling might also suggest possible clusters of regions where collaboration might be effective—and to help structure federal coordination.

Equally important, reflecting on how these patchwork patterns occur could be applicable for bettering understand of the spread of other infectious diseases—such as HIV, Lyme disease, and chlamydia, or noninfectious diseases—such as diabetes, hypertension, and Alzheimer's, which would have a significant impact to spatial epidemiology and public health research (Hanson et al. 2003; Beale et al. 2008). These ideas might be extended to our understanding of public health impacts of climate change and natural disasters (heat-related illness and death, air-quality impact on comorbidities, flooding, disasters), and how cities and regions have stepped forward in the decision-making arena in the wake of federal/national divestment, denial, or impasse. This exploratory research points to an opportunity to collaboratively work at codeveloping models that better inform local decision makers, where the goal may not be to replicate or reproduce the models, but generalize the knowledge produced to be used locally (Kedron et al. 2019). Our findings may prompt ideas on holding officials accountable to public health outcomes with their decisions—across all of these intersecting complex, mismatched scales, and reinforce the calls for future national-scale coordination with an attention to spatial congruence.

NOTES

1 The choice of Arizona is somewhat arbitrary, as any state could serve the purpose to illustrate the patchwork character of model performance. Arizona has the distinction of being among the top three states of earliest confirmed COVID-19 cases, with the longest time lag to a first decision to mediate.

2 While contributing to our process of developing the mathematical models as noted in the following sections, these decision data themselves were ultimately not directly included in the models as presented here. The specific proclamations of gubernatorial decisions are not rapidly time-varying factors, which makes using them in such models ineffective, so the more temporally variable factors accounted for the impact of the decisions on mobility data and mask data, for example.

ACKNOWLEDGMENTS

Thanks to Katsiaryna Varfalameyeva for analytical support. We appreciate the feedback on early conceptualization from Dr. Ariane Middel.

FUNDING

This research received funding from the US National Science Foundation, Award No. 2029044, RAPID: Active Tracking of Disease Spread in COVID-19 via Graph Predictive Analytics, Gautam Dasarathy, Principal Investigator. The ASU Knowledge Exchange for Resilience is supported by Virginia G. Piper Charitable Trust. Piper Trust supports organizations that enrich health, well-being, and opportunity for the people of Maricopa County, Arizona.

REFERENCES

Alexander, M. J., M. V. Kiang, and M. Barbieri. 2018. Trends in Black and White Opioid Mortality in the United States, 1979–2015. *Epidemiology* 295: 707–715. 10.1097/ede.0000000000000858.
Altekruse, S. F., C. M. Cosgrove, W. C. Altekruse, R. A. Jenkins, and C. Blanco. 2020. Socioeconomic Risk Factors for Fatal Opioid Overdoses in the United States: Findings from

the Mortality Disparities in American Communities Study (MDAC). Edited by Becky L. Genberg. *PLOS ONE* 151: e0227966. 10.1371/journal.pone.0227966.

Anastassopoulou, C., L. Russo, A. Tsakris, and C. Siettos. 2020. Data-Based Analysis, Modelling and Forecasting of the COVID-19 Outbreak. Edited by Sreekumar Othumpangat. *PLOS ONE* 153: e0230405. 10.1371/journal.pone.0230405.

Beale, L., J. J. Abellan, S. Hodgson, and L. Jarup. 2008. Methodologic Issues and Approaches to Spatial Epidemiology. *Environmental Health Perspectives* 1168: 1105–1110. 10.1289/ehp.10816.

Box, G. E. P., G. M. Jenkins, and G. C. Reinsel. 2011. *Time Series Analysis: Forecasting and Control.* New York: John Wiley & Sons.

Bozdogan, H. 1987. Model Selection and Akaike's Information Criterion (AIC): The General Theory and Its Analytical Extensions. *Psychometrika* 523: 345–370. 10.1007/BF02294361.

Butler, D. 2009. Swine Flu Goes Global. *Nature* 4587242: 1082–1083. 10.1038/4581082a.

Campbell, C. I., C. Weisner, L. LeResche, G. Thomas Ray, K. Saunders, M. D. Sullivan, and C. J. Banta-Green, et al. 2010. Age and Gender Trends in Long-Term Opioid Analgesic Use for Noncancer Pain. *American Journal of Public Health* 100(12): 2541–2547. 10.2105/AJPH.2009.180646.

Center for Disease Control [CDC]. 2014. Enhanced Ebola Screening to Start at Five U.S. Airports and New Tracking Program for All People Entering United States from Ebola-Affected Countries Washington, D.C. https://www.cdc.gov/media/releases/2014/p1008-ebola-screening.html

――――. 2020. CDC COVID Data Tracker. Washington, D.C. https://covid.cdc.gov/covid-data-tracker/

Center, P. R. 2020. Most Approve of National Response to COVID-19 in 14 Advanced Economies. *Pew Research Center* 23.

Ceylan, Z. 2020. Estimation of COVID-19 Prevalence in Italy, Spain, and France. *Science of the Total Environment* 138817April: 138817. 10.1016/j.scitotenv.2020.138817.

Chakraborty, T., and I. Ghosh. 2020. Real-Time Forecasts and Risk Assessment of Novel Coronavirus (COVID-19) Cases: A Data-Driven Analysis. *Chaos, Solitons, and Fractals* 109850 135: 109850. 10.1016/j.chaos.2020.109850.

Chakravarthy, B., S. Shah, and S. Lotfipour. 2011. Prescription Drug Monitoring Programs and Other Interventions to Combat Prescription Opioid Abuse. *Western Journal of Emergency Medicine* 135: 422–425. 10.5811/westjem.2012.7.12936.

Choo, E. K., C. Douriez, and T. Green. 2014. Gender and Prescription Opioid Misuse in the Emergency Department. Edited by Mark Mycyk. *Academic Emergency Medicine* 2112: 1493–1498. 10.1111/acem.12547.

Coker, R. 2009. Swine Flu. *BMJ* 3383: b1791–91. 10.1136/bmj.b1791.

Ding, M., Y. Chen, and S. L. Bressler. 2006. 17 Granger Causality: Basic Theory and Application to Neuroscience. In *Handbook of Time Series Analysis: Recent Theoretical Developments and Applications,* 437. Wiley Online Library.

Djulbegovic, B., D. J. Weiss, and I. Hozo. 2020. Evaluation of the U.S. Governors' Decision When to Issue Stay-At-Home Orders. *Journal of Evaluation in Clinical Practice* 265: 1347–1351. 10.1111/jep.13458.

Fineberg, H. V. 2014. Pandemic Preparedness and Response — Lessons from the H1N1 Influenza of 2009. *New England Journal of Medicine* 370(14): 1335–1342. 10.1056/nejmra1208802.

Fotheringham, A. S., W. Yang, and W. Kang. 2017. Multiscale Geographically Weighted Regression (MGWR). *Annals of the American Association of Geographers* 1076: 1247–1265. 10.1080/24694452.2017.1352480.

Gehlke, C. E., and K. Biehl. 1934. Certain Effects of Grouping upon the Size of the Correlation Coefficient in Census Tract Material. *Journal of the American Statistical Association* 29185: 169. 10.2307/2277827.

Gibbs, W. W., and C. Soares. 2005. Preparing for a Pandemic. *Scientific American* 2935: 44–54. 10.1038/scientificamerican1105-44.

Gostin, L. O., J. G. Hodge, and S. Burris. 2014. Is the United States Prepared for Ebola? *JAMA* 31223: 2497. 10.1001/jama.2014.15041.

Gostin, L. O., J. G. Hodge, and L. F. Wiley. 2020. Presidential Powers and Response to COVID-19. *JAMA* 32316: 1547–1548. 10.1001/jama.2020.4335.

Graziani, M., and R. Nisticò. 2016. Gender Difference in Prescription Opioid Abuse: A Focus on Oxycodone and Hydrocodone. *Pharmacological Research* 108: 31–38. 10.1016/j.phrs.2016.04.012. June.

Haffajee, R. L., and M. M. Mello. 2020. Thinking Globally, Acting Locally—The U.S. Response to COVID-19. *New England Journal of Medicine* 382: 22. 10.1056/nejmp2006740.

Hale, T., N. Angrist, R. Goldszmidt, B. Kira, A. Petherick, T. Phillips, and S. Webster. others. 2021. A Global Panel Database of Pandemic Policies (Oxford COVID-19 Government Response Tracker). *Nature Human Behaviour* 1–10. 10.1038/s41562-021-01079-8. March.

Hanage, W. P., C. Testa, J. T. Chen, L. Davis, E. Pechter, M. Santillana, and N. Krieger. 2020. COVID-19: U.S. Federal Accountability for Entry, Spread, and Inequities. *Harvard Center for Population and Development Studies* 20(2). [https://cdn1.sph.harvard.edu/wp-content/uploads/sites/1266/2020/10/20_covid-19_federal-response_FINAL_for-HCPDS_1001_HCPDS-working-paper_volume-20_number-2_FINAL.pdf.

Hanson, K., M. K. Ranson, V. Oliveira-Cruz, and A. Mills. 2003. Expanding Access to Priority Health Interventions: A Framework for Understanding the Constraints to Scaling-up. *Journal of International Development* 151: 1–14. 10.1002/jid.963.

Holguín-Veras, J., N. Pérez, S. Ukkusuri, T. Wachtendorf, and B. Brown. 2007. Emergency Logistics Issues Affecting the Response to Katrina. *Transportation Research Record: Journal of the Transportation Research Board* 20221: 76–82. 10.3141/2022-09.

Jacobson, P. D., D. Chrysler, and J. Bresler. 2020. Executive Decision Making for COVID-19: Public Health Science through a Political Lens. Assessing Legal Responses to COVID-19, August, 7.

Kane, M. J., N. Price, M. Scotch, and P. Rabinowitz. 2014. Comparison of ARIMA and Random Forest Time Series Models for Prediction of Avian Influenza H5N1 Outbreaks. *BMC Bioinformatics* 15: 1. 10.1186/1471-2105-15-276.

Kang, D., H. Choi, J.-H. Kim, and J. Choi. 2020. Spatial Epidemic Dynamics of the COVID-19 Outbreak in China. *International Journal of Infectious Diseases*, 94: 96–102. 10.1016/j.ijid.2020.03.076.

Kedron, P., A. E. Frazier, A. B. Trgovac, T. Nelson, and A. S. Fotheringham. 2019. Reproducibility and Replicability in Geographical Analysis. *Geographical Analysis* 531: 135–147. 10.1111/gean.12221.

Kettl, D. F. 2006. Is the Worst yet to Come? *The ANNALS of the American Academy of Political and Social Science* 6041: 273–287. 10.1177/0002716205285981.

Keyes, K. M., M. Cerdá, J. E. Brady, J. R. Havens, and S. Galea. 2014. Understanding the Rural–Urban Differences in Nonmedical Prescription Opioid Use and Abuse in the United States. *American Journal of Public Health* 1042: e52–59. 10.2105/ajph.2013.301709.

Krumkamp, R., H.-P. Duerr, R. Reintjes, A. Ahmad, A. Kassen, and M. Eichner. 2009. Impact of Public Health Interventions in Controlling the Spread of SARS: Modelling of Intervention Scenarios. *International Journal of Hygiene and Environmental Health* 2121: 67–75. 10.1016/j.ijheh.2008.01.004.

Lagu, T., A. W. Artenstein, and R. M. Werner. 2020. Fool Me Twice: The Role for Hospitals and Health Systems in Fixing the Broken PPE Supply Chain. *Journal of Hospital Medicine* 159: 570–571. 10.12788/jhm.3489.

Leatherby, L., and R. Harris. 2020. States that Imposed Few Restrictions Now Have the Worst Outbreaks. *The New York Times*, November 18.

Leitner, H., E. Sheppard, and K. M. Sziarto. 2008. The Spatialities of Contentious Politics. *Transactions of the Institute of British Geographers* 332: 157–172. 10.1111/j.1475-5661.2008.00293.x.

Leung, G. M., and A. Nicoll. 2010. Reflections on Pandemic (H1N1) 2009 and the International Response. *PLoS Medicine* 710: e1000346. 10.1371/journal.pmed.1000346.

Li, H., H. Li, Z. Ding, Z. Hu, F. Chen, K. Wang, Z. Peng, and H. Shen. 2020. Spatial Statistical Analysis of Coronavirus Disease 2019 (COVID-19) in China. *Geospatial Health* 15: 1. 10.4081/gh.2020.867.

Li, Z., and A. S. Fotheringham. 2020. Computational Improvements to Multi-Scale Geographically Weighted Regression. *International Journal of Geographical Information Science* 347: 1378–1397. 10.1080/13658816.2020.1720692.

Liaw, A., and M. Wiener, et al. 2002. Classification and Regression by RandomForest. *R News* 23: 18–22.

Maffioli, E. M. 2020. How Is the World Responding to the Novel Coronavirus Disease (COVID-19) Compared with the 2014 West African Ebola Epidemic? the Importance of China as a Player in the Global Economy. *The American Journal of Tropical Medicine and Hygiene* 1025: 924–925. 10.4269/ajtmh.20-0135.

Mallinson, D. J. 2020. Cooperation and Conflict in State and Local Innovation during COVID-19. *The American Review of Public Administration* 50(6–7): 543–550. 10.1177/0275074020941699.

Mollalo, A., B. Vahedi, and K. M. Rivera. 2020. GIS-Based Spatial Modeling of COVID-19 Incidence Rate in the Continental United States. *Science of the Total Environment* 728138884: 138884. 10.1016/j.scitotenv.2020.138884.

Moran, P. A. P. 1947. Random Associations on a Lattice. *Mathematical Proceedings of the Cambridge Philosophical Society* 433: 321–328. 10.1017/s0305004100023550.

———. 1948. The Interpretation of Statistical Maps. *Journal of the Royal Statistical Society: Series B (Methodological)* 10(2): 243–251. 10.1111/j.2517-6161.1948.tb00012.x.

National Academies of Sciences, Engineering, and Medicine. 2019. *Reproducibility and Replicability in Science*. Washington, D.C.: The National Academies Press. 10.17226/25303.

Nelson, L. S., D. N. Juurlink, and J. Perrone. 2015. Addressing the Opioid Epidemic. *JAMA* 31414: 1453. 10.1001/jama.2015.12397.

Openshaw, S. 1984. The Modifiable Areal Unit Problem. *Geo Abstracts* Norwich, U.K.: University of East Anglia

Orenstein, W. A., B. G. Gellin, T. Buck, L. A. Jackson, P. S. LaRussa, J. O. Mason, and M. McCormick, and others. 2013. Strategies to Achieve the Healthy People 2020 Annual Influenza Vaccine Coverage Goal for Health-Care Personnel: Recommendations from the National Vaccine Advisory Committee. *Public Health Reports* 1281: 7–25. 10.1177/003335491312800103.

Park, B. J., A. J. Peck, M. J. Kuehnert, C. Newbern, C. Smelser, J. A. Comer, D. Jernigan, and L. C. McDonald. 2004. Lack of SARS Transmission among Healthcare Workers, United States. *Emerging Infectious Diseases* 10(2): 217–224. 10.3201/eid1002.030793.

Pedregosa, F., G. Varoquaux, A. Gramfort, V. Michel, B. Thirion, O. Grisel, and M. Blondel, et al. 2011. Scikit-Learn: Machine Learning in Python. *The Journal of Machine Learning Research* 12 JMLR. org: 2825–2830.

Pikoulis, E., K. Puchner, E. Riza, E. Kakalou, E. Pavlopoulos, C. Tsiamis, V. Tokakis, G. Boustras, A. Terzidis, and V. Karamagioli. 2020. In the Midst of the Perfect Storm: Swift Public Health Actions Needed in order to Increase Societal Safety during the COVID-19 Pandemic. *Safety Science* 129104810: 104810. 10.1016/j.ssci.2020.104810.

Pletcher, M. J., S. G. Kertesz, M. A. Kohn, and R. Gonzales. 2008. Trends in Opioid Prescribing by Race/Ethnicity for Patients Seeking Care in US Emergency Departments. *JAMA* 299: 1. 10.1001/jama.2007.64.

Praharaj, S., D. King, C. Pettit, and E. A. Wentz. 2020. Using Aggregated Mobility Data to Measure the Effect of COVID-19 Policies on Mobility Changes in Sydney, London, Phoenix, and Pune. *Findings*, 20 October. 10.32866/001c.17590

Quinn, S. C. 2006. Hurricane Katrina: A Social and Public Health Disaster. *American Journal of Public Health* 962: 204–4. 10.2105/ajph.2005.080119.

Reibman, J., N. Levy-Carrick, T. Miles, K. Flynn, C. Hughes, M. Crane, and R. G. Lucchini. 2016. Destruction of the World Trade Center Towers. Lessons Learned from an Environmental Health Disaster. *Annals of the American Thoracic Society* 135: 577–583. 10.1513/AnnalsATS.201509-572PS.

Reiner, R. C., R. M. Barber, J. K. Collins, P. Zheng, C. Adolph, J. Albright, and C. M. Antony, et al. 2020. Modeling COVID-19 Scenarios for the United States. *Nature Medicine* October: 1–12. 10.1038/s41591-020-1132-9.

Rigg, K. K., and S. M. Monnat. 2015. Urban Vs. Rural Differences in Prescription Opioid Misuse among Adults in the United States: Informing Region Specific Drug Policies and Interventions. *International Journal of Drug Policy* 265: 484–491. 10.1016/j.drugpo.2014.10.001.

Rosenfield, A., S. S. Morse, and K. Yanda. 2002. September 11: The Response and Role of Public Health. *American Journal of Public Health* 921: 10–11. 10.2105/ajph.92.1.10.

Rothstein, M. A. 2015. From SARS to Ebola: Legal and Ethical Considerations for Modern Quarantine. *Indiana Health Law Review* 121: 227–280. 10.18060/18963.

Schneider, S. K. 2005. Administrative Breakdowns in the Governmental Response to Hurricane Katrina. *Public Administration Review* 655: 515–516. 10.1111/j.1540-6210.2005.00478.x.

Smith, R. D. 2006. Responding to Global Infectious Disease Outbreaks: Lessons from SARS on the Role of Risk Perception, Communication and Management. *Social Science & Medicine* 6312: 3113–3123. 10.1016/j.socscimed.2006.08.004.

Sobel, R. S., and P. T. Leeson. 2006. Government's Response to Hurricane Katrina: A Public Choice Analysis. *Public Choice* 1271–2: 55–73. 10.1007/s11127-006-7730-3.

Solís, P., J. Vanos, and R. Forbis. 2017. The Decision-making/Accountability Spatial Incongruence Problem for Research Linking Science and Policy. *The Geographical Review* 1074: 680–704. 10.1111/gere.12240.

SteelFisher, G. K., R. J. Blendon, and N. Lasala-Blanco. 2015. Ebola in the United States—Public Reactions and Implications. *New England Journal of Medicine* 3739: 789–791. 10.1056/nejmp1506290.

The White House, 2020. The President's Coronavirus Guidelines for America. Washington, D.C. 780 https://www.whitehouse.gov/wp-content/uploads/2020/03/03.16.20_coronavirus-guidance_8.5x11_315PM.pdf

Thomson, A., G. Vallée-Tourangeau, and L. S. Suggs. 2018. Strategies to Increase Vaccine Acceptance and Uptake: From Behavioral Insights to Context-Specific, Culturally-Appropriate, Evidence-Based Communications and Interventions. *Vaccine* 3644: 6457–6458. 10.1016/j.vaccine.2018.08.031.

Tobler, W. R. 1970. A Computer Movie Simulating Urban Growth in the Detroit Region. *Economic Geography* 46: June 234. 10.2307/143141.

U.S. Department of Health and Human Services. 2020. Determination that a Public Health Emergency Exists. *Public Health Emergency.* January 31. Washington, D.C. https://www.phe.gov/emergency/news/healthactions/phe/Pages/2019-nCoV.aspx

Wang, C., Z. Li, M. Mathews, S. Praharaj, B. Karna, and P. Solís. 2020a. The Spatial Association of Social Vulnerability with COVID-19 Prevalence in the Contiguous United States. *International Journal of Environmental Health Research* 10.1080/09603123.2020.1847258. 1–8.

Wang, C., L. Liu, X. Hao, H. Guo, Q. Wang, J. Huang, and N. He, et al. 2020b. Evolving Epidemiology and Impact of Non-Pharmaceutical Interventions on the Outbreak of Coronavirus Disease 2019 in Wuhan, China. *MedRxiv.* Cold Spring Harbor, N.Y.: Cold Spring Harbor Laboratory Press.

Wang, P., X. Zheng, J. Li, and B. Zhu. 2020c. Prediction of Epidemic Trends in COVID-19 with Logistic Model and Machine Learning Technics. *Chaos, Solitons, and Fractals* 139:110058. Elsevier. 10.1016/j.chaos.2020.110058.

Warren, M. S., and S. W. Skillman. 2020. Mobility Changes in Response to COVID-19. ArXiv Preprint ArXiv:2003.14228.

Wils, F. 1996. Scaling Up, Mainstreaming, and Accountability: The Challenge to NGOs. In *Beyond the Magic Bullet: NGO Performance and Accountability in the Post-Cold War World*, edited by M. Edwards and D. Hulme, 67–79. West Hartford, Conn.: Kumarian Press.

Wu, J. T., K. Leung, and G. M. Leung. 2020. Nowcasting and Forecasting the Potential Domestic and International Spread of the 2019—NCoV Outbreak Originating in Wuhan, China: A Modelling Study. *The Lancet* 395: 10225. 10.1016/s0140-6736(20)30260-9.

Yang, Z., Z. Zeng, K. Wang, -S.-S. Wong, W. Liang, M. Zanin, and P. Liu, et al. 2020. Modified SEIR and AI Prediction of the Epidemics Trend of COVID-19 in China under Public Health Interventions. *Journal of Thoracic Disease* 123: 165–174. 10.21037/jtd.2020.02.64.

Yong, E. 2020. We All Live in a Patchwork Pandemic Now. *The Atlantic.* 20 May. https://www.theatlantic.com/health/archive/2020/05/patchwork-pandemic-states-reopening-inequalities/611866/20May

Zhang, S., M. Y. Diao, W. Yu, L. Pei, Z. Lin, and D. Chen. 2020. Estimation of the Reproductive Number of Novel Coronavirus (COVID-19) and the Probable Outbreak Size on the Diamond Princess Cruise Ship: A Data-Driven Analysis. *International Journal of Infectious Diseases* 93: 201–204. 10.1016/j.ijid.2020.02.033.

Index

Note: Figures are indicated by *italics*. Tables are indicated by **bold**. Endnotes are indicated by the page number followed by 'n' and the endnote number e.g., 20n1 refers to endnote 1 on page 20.

For Product Safety Concerns and Information please contact our EU
representative GPSR@taylorandfrancis.com
Taylor & Francis Verlag GmbH, Kaufingerstraße 24, 80331 München, Germany

9 781032 447148